Algorithmic Foundation of
Multi-Scale Spatial Representation

Algorithmic Foundation of
Multi-Scale Spatial Representation

Zhilin Li

Department of Land Surveying and Geo-Informatics
The Hong Kong Polytechnic University

CRC Press
Taylor & Francis Group
Boca Raton London New York

CRC Press is an imprint of the
Taylor & Francis Group, an **informa** business

CRC Press
Taylor & Francis Group
6000 Broken Sound Parkway NW, Suite 300
Boca Raton, FL 33487-2742

First issued in paperback 2020

© 2007 by Taylor & Francis Group, LLC
CRC Press is an imprint of Taylor & Francis Group, an Informa business

No claim to original U.S. Government works

ISBN-13: 978-0-367-57779-7 (pbk)
ISBN-13: 978-0-8493-9072-2 (hbk)

Library of Congress Cataloging-in-Publication Data

Li, Zhilin.
   Algorithmic foundation of multi-scale spatial representation / by Zhilin Li.
     p. cm.
   ISBN 0-8493-9072-9
   1. Geographic information systems. 2. Algorithms. I. Title.

G70.212L487 2006
526.801'5181--dc22
                                          2006045503

Visit the Taylor & Francis Web site at
http://www.taylorandfrancis.com

and the CRC Press Web site at
http://www.crcpress.com

Preface

The representation of spatial data, or simply spatial representation, can be made in various forms, that is, graphics or nongraphics. *Graphic representation* includes maps, images, drawings, diagrams, movies, and animation, either 2-dimensional (2-D) or 3-D. *Nongraphical representation* includes audio, text, digital numbers, and so on. Graphic representation is still the more popular alternative in geo-information science and thus will be the focus of this book. Spatial data can also be represented at different scales, leading to the issues of *multi-scale representation*, which is the topic of this book.

Multi-scale representation of spatial data is termed *multi-scale spatial representation* in this context. It is simply termed *multiple representation* in some literature. Multi-scale representation is a traditional topic in cartography, geography, and all other geo-sciences. It has also become one of the most important issues in geo-information science.

Among the many forms of graphic representation of spatial data, maps are still the most effective and popular means. The multi-scale issue related to map production is the derivation of maps at a smaller scale than those at a larger scale, which is a process traditionally termed *generalization*. *Map generalization* is also required for real-time zooming in and out in a geographic information system. Therefore, the emphasis of this book is on map generalization.

Generalization is a process of abstraction. The terrain features represented on maps at smaller scales are the abstractive representations of those at larger scales. For example, at 1:1,000,000, a city is represented by an abstractive symbol — a point symbol — while at 1:1,000, every street is represented by double lines. Perhaps the highest level of generalization was achieved by Sir Isaac Newton, who represented any planet by a point in his famous "Law of Gravitation." This means that generalization is not only an issue related to the representation of spatial data but also to the modeling of spatial processes.

Although map generalization is a traditional topic, the generalization of maps in a digital environment, simply digital map generalization, is a result of computerization in recent decades. Although it is difficult to locate the exact date at which research on digital map generalization started, three very important publications appeared in 1966: Tobler (1966), Töpfer and Pillewizer (1966), and Perkal (1966). From 1966 to the early 1980s, research on digital map generalization was rather isolated. In the 1970s, efforts were made to develop algorithms for line features. In the early 1980s, those line algorithms were evaluated and algorithms for area features were investigated. In the later 1980s, more attempts were made to develop strategies and rule-based systems.

As early as the 1980s, multi-scale representation was recognized as a fundamental issue in spatial data handling. In 1983, a small group of leading scientists in the United States was gathered by NASA to define critical research areas in spatial data handling, and multi-scale representation was identified as one of them (Marble, 1984). Since then, this topic has become part of the international research agenda in spatial information sciences. Indeed, many researchers have advocated research on this topic (Abler, 1987; Rhind, 1988; Müller, 1990). As a result, the topic has attracted the attention of researchers in relevant disciplines, and a number of research projects on this topic have been initiated internationally. The importance of multi-scale representation was highlighted by an NCGIA (National Center for Geographic Information and Analysis) initiative under the title "Multiple Representation" in 1989.

Since the early 1990s, multi-scale representation (or generalization) has been a popular research topic, with the widespread use of the Geographic Information System (GIS), which integrates multi-scale, multi-source spatial data. Multi-scale representation has also become an important topic in computational geometry and computing. The International Cartographic Association (ICA) established a working group on automated map generalization in 1991, which became a commission in 2001. This commission, chaired by Professor Robert Weibel (University of Zurich) until 2003, has made significant advancement. In 2000, the International Society for Photogrammetry and Remote Sensing (ISPRS) also established a working group on multi-scale representation. Special sessions on this topic have been organized at conferences, and special issues have been published in journals. Indeed, multi-scale representation gained a boost in the later 1990s with the funding of the AGENT project by the European Union.

Forty years have passed since the publication of the first three important papers in 1966. An age of 40 has special meaning in Chinese: *enlightenment.* It means one will rarely be puzzled after this age. Accordingly, a discipline after 40 years of development should be well established. This maturity is signified by the growing number of publications in the discipline. We have seen two edited books (resulting from expert meetings) (Buttenfield and McMaster, 1991; Müller et al., 1995) and a Ph.D. thesis (João, 1998) published by formal publishers. We have also seen a resource booklet (McMaster and Shea, 1992) published by The Association of American Geographers. Unfortunately, no book has been authored to systematically address different aspects of the discipline. The author of this book feels it is extremely difficult to cover all the aspects of multi-scale spatial representation in a single book at the present time. Therefore, a compromise is struck here to provide a book only addressing the mathematical basis of the discipline, more precisely the algorithmic foundation.

In multi-scale spatial representation, various types of transformations are undertaken, that is, geometric, thematic, and topological transformations. In this book, only geometric transformations are discussed. Indeed, this book covers the low-level algorithms available for the multi-scale representations of different types of spatial features, that is, point clusters, individual line features, a class of line features (contours, transportation networks, and hydrographic networks), individual area features, and a class of area features. In addition, algorithms for multi-scale representation of 3-D surfaces and 3-D features are briefly discussed.

This book consists of 12 chapters. Chapter 1 is the introduction, providing an overview of the contents. Chapters 2 and 3 provide some mathematical and theoretical foundations to facilitate the discussions in the later chapters. Chapter 4 describes the algorithms for a class of point features (or point clusters). Chapters 5 through 7 are devoted to algorithms for individual line features — Chapter 5 for reduction of data points, Chapter 6 for smoothing (filtering), and Chapter 7 for scale-driven generalization. Chapter 8 discusses the algorithms for a set of line features, namely, contour set, river network, and transportation network. Chapter 9 addresses the algorithms for individual area features, and Chapter 10 discusses a set of area features. Chapter 11 covers the algorithms for various displacement operations. Chapter 12 provides brief coverage of the algorithms for 3-D surfaces and 3-D features.

Indeed, this book provides comprehensive coverage of the algorithmic foundation of multi-scale representation of spatial data. It is written at a medium level of technical detail to make the concepts easy to understand. An attempt is also made to use illustrations, to make the working principles of algorithms intuitive.

REFERENCES

Abler, R., The National Science Foundation National Center for Geographc Information and Analysis. *International Journal of Geographical Information Systems,* 1(4), 303–326, 1987.

Buttenfield, B. P. and McMaster R. B., Eds., *Map Generalization: Making Rules for Knowledge Representation*, Longman Scientific and Technical, London, 1991.

João, E., *Causes and Consequences of Map Generalization.* CRC Press, Boca Raton, 1998.

Marble, D., Geographic information systems: an overview. *Proceedings of Pecora 9*, Sioux Falls, SD, 1984, 18–24. Reprinted in Peuquet, D. J. and Marble, D. F. Eds., *Introductory Readings in Geographic Information Systems*, Taylor & Francis, London, 1990, pp. 8–17.

McMaster, R. B. and Shea, K. S., *Generalization in Digital Cartography*, Association of American Geographers, Washington, DC, 1992.

Müller, J.-C., Rule based generalization: potentials and impediments, in *Proceedings of 4th International Symposium on Spatial Data Handling,* International Geographic Union, 1990, pp. 317–334.

Müller, J. -C., Lagrange, J. P., and Weibel, R., Eds., *GIS And Generalization: Methodology and Practice*, Taylor & Francis, London, UK, 1995.

Perkal, J. D., An Attempt at Objective Generalisation (translated by W. Jackowshi from Proba obiektywnej generalizacji, Geodezia I Kartografia, Tom VII, Zeszyt 2, 1958, pp. 130–142), Michigan Inter-University Community of Mathematical Geographers Discussion Paper 10, Dept. of Geography, University of Michigan, Ann Arbor, MI, 1966.

Rhind, D., A GIS research agenda. *International Journal of Geographical Information Systems,* 2(1), 23–28, 1988.

Tobler, W., *Numerical map generalization.* Michigan Inter-University Community of Mathematical Geographers Discussion Paper 8. Dept. of Geography, University of Michigan, An Arbor, MI, USA, 1966.

Töpfer, F. and W. Pillewizer, The principles of selection, *Cartographic Journal*, 3(1), 10–16, 1966.

Acknowledgments

First, I would like to take this opportunity to express my sincere gratitude to Prof. Stan Openshaw for allowing me to enter into this area with great freedom. Indeed, it was with Stan at the NorthEast Regional Research Laboratory (NE.RRL) at The University of Newcastle upon Tyne that I was for the first time attracted to this topic. That was really an enjoyable period (1990 to 1991) because the ESRC (Environmental and Social Research Council) of the United Kingdom supported three research associates for each of its eight regional research laboratories (RRL) and gave the RRLs great freedom to select research topics.

I would also like to express my thanks to fellow colleagues in the community for their constructive discussions on map generalization research, especially Prof. David Rhind, Prof. Jean-Claude Muller, Prof. Robert Weibel, Dr. Bo Su, Prof. Liqiud Meng, Prof. Robert Bob McMaster, Prof. Barbara Buttenfield, Prof. Deren Li, Prof. Jiayao Wang, Prof. Yue Liu, Prof. Jun Chen, Prof. Tinghua Ai, Prof. Haowen Yan, Dr. Zesheng Wang, Dr. Dan Lee, Dr. Elsa João, and Dr. Tiina Sarjakoski. My sincere appreciation goes to T. Ai, T. Cheng, M. Deng, H. Qi, and Y. Hu for their constructive comments on the early versions of this book and to the publisher for making this volume available to you.

I am in debt to the National Natural Science Foundation of China and the Research Grant Council (RGC) of Hong Kong Special Administration Region for their support through a number of research grants.

I gratefully acknowledge the kind permissions granted by a number of organizations and individuals that have allowed me to make use of their copyrighted materials.

Last but not least, special thanks go to my wife, Lingyun Liu for her continuous support and understanding and to our sons, Andrew and Edward for their help in the writing of this book.

Author

Dr. Zhilin Li is a full professor of geo-informatics (cartography/GIS/remote sensing) at the Department of Land Surveying and Geo-Informatics, the Hong Kong Polytechnic University. He holds a B.Eng. and Ph.D. Since obtaining his Ph.D. from the University of Glasgow (U.K.) in 1990, he has worked as a research fellow at the University of Newcastle upon Tyne, (U.K.), the University of Southampton (U.K.), and the Technical University of Berlin (Germany). He had also worked at Curtin University of Technology (Australia) as a lecturer for two years. He joined the Hong Kong Polytechnic University in early 1996.

Prof. Li has published 90 papers in international journals and is the principal author of the popular book *Digital Terrain Modeling: Principles and Methodology*. He has been presented with the *Schwidefsky Medal* by the International Society for Photogrammetry and Remote Sensing (ISPRS) at its 20th Congress held in 2004 and the *State Natural Science Award* from the Central Government of China in 2005. Prof. Li's research interests include multi-scale spatial representation and map generaliazation, digital terrain and 3-D modelling, and spatial relations.

Contents

1 Introduction

Spatial representation refers to the representation of spatial data in this context. Spatial data could be represented at different scales, leading to the issues of *multi-scale spatial representation*. This chapter provides an overview of the issues of spatial representation and multi-scale spatial representation. Emphasis is on the introduction of the essential operations for geometric transformations in multi-scale spatial representation.

1.1 SPATIAL REPRESENTATION: REPRESENTATION OF SPATIAL DATA

1.1.1 FORMS OF SPATIAL REPRESENTATION

Spatial data can be represented in various forms, that is, graphics or nongraphics. *Graphic representation* includes maps, drawings, and animation, either two-dimensional (2-D) or 3-D. Graphic representations are achieved through the use of point, linear, and areal symbols. Colors and patterns are also in use as visual variables. Nongraphical representation includes audio, audiovisual, text, digital numbers, and so on. Nongraphical representations are not very popular in the geo-information community and thus are not discussed in this book.

A *map* is a typical 2-D representation. It is a spatial representation of reality, used for recording and conveying information about spatial and semantic characteristics of the natural world and cultural phenomena. It is a traditional means for spatial representation. Figure 1.1 shows a spatial representation on an ancient Chinese map (made in the Han Dynasty before 168 BC). Maps have been a popular mode of spatial representation and are still popularly used in practice because of their measurability, which results from the use of mathematical laws, intuitive view from symbolization, and overview from generalization.

Based on their contents, maps are usually classified into three types:

- Topographic maps (also called general-purpose maps): Representation of terrain surface and the features on the surface with balance (Figure 1.2a). (See color insert following page 116).
- Thematic maps: Representation of a theme of natural or cultural phenomena (Figure 1.2b).
- Special maps: Representation of a few themes of natural and/or cultural phenomena, for example, tourist maps. In other words, they are in between topographic and thematic maps.

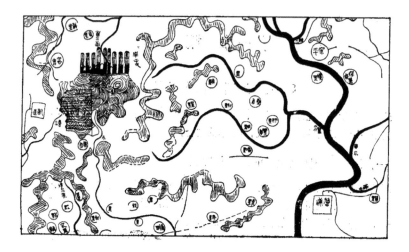

FIGURE 1.1 Representation of spatial data on an ancient Chinese map.

A topographic map is a type of qualitative map, but a thematic map could be either quantitative or qualitative. Other criteria can also be used for classification of maps such as scale, usage, size of area covered, color, and so on. One interesting classification made by Moellering (1984, 1987) distinguishes maps into:

- Real maps: Directly viewable and having a permanent tangible reality.
- Virtual maps: Lacking one or both of the qualities of real maps.

Moellering (1984, 1987) has further differentiated virtual maps into three types:

- Virtual type 1: Directly viewable but without a permanent tangible reality, for example, a screen display.
- Virtual type 2: Having a permanent tangible reality but not directly viewable, for example, hardcopy form on a CD-ROM.
- Virtual type 3: Neither hardcopy nor viewable, for example, data stored on disk.

(a) Topographic map at 1:200,000   (b) Thematic map (house development)

FIGURE 1.2 (See color insert following page 116) A topographic map (a) and a thematic map (b) (Courtesy of LIC of HKSAR).

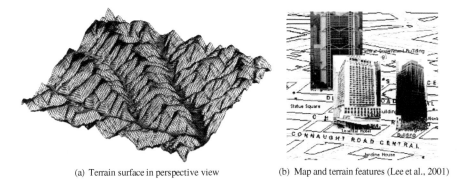

(a) Terrain surface in perspective view   (b) Map and terrain features (Lee et al., 2001)

FIGURE 1.3 Three-dimensional representation of terrain surface and terrain features.

With the advancement of computing technology, *3-D representation* has become more and more popular. Spatial data can be represented in perspective views. Figure 1.3a is an example generated from a digital terrain model, or digital elevation model (see Li et al., 2005). *Rendering techniques* can be applied to produce more realistic representations in image form. Terrain texture and features can also be added to perspective views (Figure 1.3b) or rendered representations to generate more vivid representations. These are not true 3-D and are sometimes called $2^1/_2$-D representations.

A true 3-D representation is viewed stereoscopically. Each stereo view consists of two graphical representations (or images), one viewed by each eye. Stereo viewing is a common technique for increasing visual realism or enhancing user interaction with 3-D scenes. The 3-D effect can be created by introducing a visual parallax along the eye-base direction, that is, the x-direction in the conventional coordinate system. The principle of such a representation lies outside the scope of this book and can be found elsewhere (e.g., Li et al., 2005). Figure 1.4 shows a pair of contour maps with a stereo view. A special viewing device is required to force the left eye to view the left image (or spatial representation) and the right eye to view the right one. Optical or optical–mechanical devices are required for this purpose. Figure 1.5 shows two examples of such optical–mechanical devices.

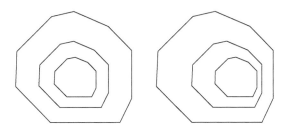

FIGURE 1.4 Visual parallax introduced into the contour map to create a stereo view.

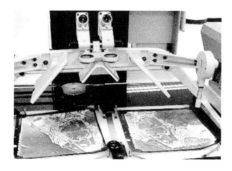

FIGURE 1.5 Two examples of optical–mechanical devices for stereo viewing.

1.1.2 DYNAMICS OF SPATIAL REPRESENTATION

Spatial data can be represented either in static or in dynamic mode. Traditional maps are typical examples of *static spatial representation*. Indeed, those discussed in Section 1.1.1 are all static representations. In this context, *dynamic representation* means that some kind of interaction or movement is involved. This classification is slightly different from that made by other researchers (e.g., Kraak, 2001), and the inclusion of interaction might be arguable.

In *interactive representation*, the information for representation is controlled by the operator through some action. A typical example is the *click* (or *double-click*) operation. By clicking a symbol, more information about the feature will be displayed either from the same source or through a hyperlink. An interesting development is the use of the mouse to control the interaction between the legend and symbol. One could display a type of feature by clicking the feature symbol in the legend (as control-panel) or alternatively show the legend by clicking the feature inside the map. Another interesting development is the *mouse-over* operation. When the mouse is moved over a symbol, more information about the feature is displayed. Figure 1.6 (See color insert following page 116) shows two examples of such developments (van den Worm, 2001).

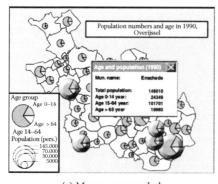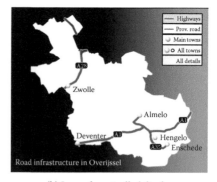

(a) Mouse-over symbol   (b) Legend-controlled display

FIGURE 1.6 (See color insert following page 116) New developments in interactive representations (van den Worm, 2001).

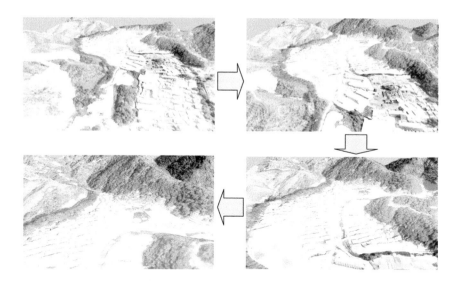

FIGURE 1.7 Four frames of a fly-through animation for a piece of terrain surface (Reprinted from Li et al., 2005).

Movement can be achieved in various ways. *Drag* and *pan* are the two simple operations to move a representation around a screen. The former moves a representation on a screen by holding the click and then moving the mouse, while the latter makes use of the scroll bars. These two operations do not change the viewpoint of the representation. The *fly-through* and *walk-through* operations create representations viewed from different positions. The former mimics the representation viewed by a bird when flying over an area, and the latter mimics the representation viewed by a person while walking along a route. These two operations are considered a result of animation. The fundamental technique used in such an animation is the page flipping, leading to movies. In the *animation* process, a number of frames (i.e., representations) are first prepared and these frames are then played in sequence. Figure 1.7 shows four frames of a fly-through animation of terrain features. To control the animation process, three variables, called *dynamic variables*, are available (DiBiase et al., 1992):

- Duration: The time units for a scene, e.g., $\frac{1}{30}$ second (30 frames per second). If the duration is too long, the action will be jerky.
- Rate of change: The pace of animation or difference between two successive scenes. If the rate is low, slow motion can be produced. Fast motion is produced if the rate is high.
- Order: The sequence of the frames, which could be arranged according to time, position, or attributes.

Motion can also be created by other types of animation. *Blinking* is an operation at symbol level and can be achieved by animating space (location) or attributes. This is a local operation. For a whole representation, motion can also be animated over

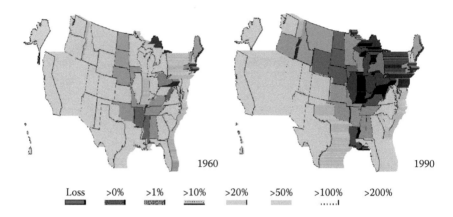

1960   1990

Loss   >0%   >1%   >10%   >20%   >50%   >100%   >200%

FIGURE 1.8 (See color insert following page 116) Two frames selected from an animation of U.S. population change since 1790.

time, scale, and attributes. Figure 1.8 (See color insert following page 116) shows two frames (i.e., 1960 and 1990) from the motion animated over time for the relative population change in the United States over time (since 1790), which were produced from IDL (interactive data language). Animation of attributes can be achieved by switching layers on and off and by reclassification of data. Animation over scale can be achieved by a *zooming* (zoom in and out) operation. Figure 1.9 (See color

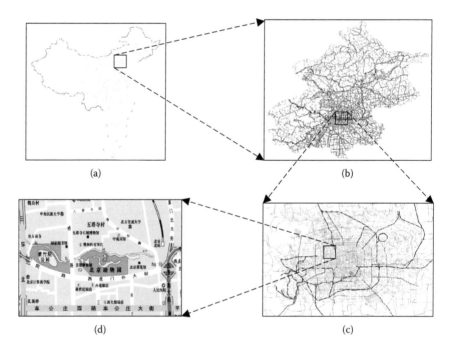

(a)   (b)

(d)   (c)

FIGURE 1.9 (See color insert following page 116) Zooming into Beijing streets as an animation (Courtesy of National Geometics Center of China).

insert following page 116) shows the continuous zooming into the Beijing streets. This is a multi-scale issue and thus is the main topic of this book. More precisely, this book introduces a set of algorithms for various operations required for multi-scale representation of spatial data.

1.2 MULTI-SCALE SPATIAL REPRESENTATION

Section 1.1.2 discussed that zooming is a process of dynamic multi-scale spatial representation and is one of the most exciting functions in a geographical information system (Abler, 1987). In the context of mapping, maps are produced at various scales, resulting in multi-scale representations in static mode. That is, multi-scale spatial representation can be in different formats and different modes. This section will discuss some general issues in multi-scale spatial representation.

1.2.1 Spatial Representation as a Record in the Scale–Time System

A spatial representation is a record of spatial phenomena at a particular time and at a particular scale in the *scale–time system* (Li, 1993), as shown in Figure 1.10a. Figure 1.10b shows a number of spatial representations at different times but at a fixed scale. This figure indicates the *transformations of spatial representation* in time. This is about the *updating* of spatial representation. Figure 1.10c shows a number of spatial representations at different scales but at the same time. This is about multi-scale representation and is referred to as *transformations in scale* in this context.

1.2.2 Transformations of Spatial Representations in Time: Updating

The environment changes over time. Natural processes (e.g., soil erosion or land subsidence) usually change slowly. Dramatic changes are caused by natural disasters and human activities. Examples are buildings being destroyed by earthquake and land lots being subdivided. The changes of coastal lines in Hong Kong over time (due to reclamation), as shown in Figure 1.11, (See color insert following page 116) is another example of human activity.

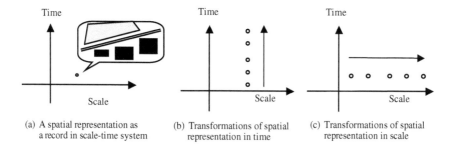

(a) A spatial representation as a record in scale-time system

(b) Transformations of spatial representation in time

(c) Transformations of spatial representation in scale

FIGURE 1.10 Spatial representations as a record in the time–scale system.

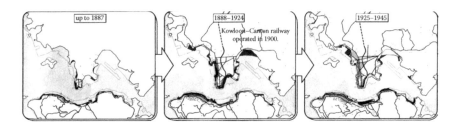

FIGURE 1.11 (See color insert following page 116) Changes of coastal lines in Hong Kong over time (Courtesy of LIC of HKSAR).

The usefulness of an outdated spatial representation (e.g., a map) is very limited for most applications except for those studies related to history. Indeed, currency (or updatedness) has been considered as a quality measure for spatial data and spatial representation in the literature (e.g., Burrough and McDonnell, 1998) and spatial data transfer standards (SDTS) (see http://mcmcweb.er.usgs.gov/sdts/). That is, spatial representations need to be updated frequently.

Updating has indeed become a headache for spatial data producers such as national mapping agencies and has become a hot research topic in recent years. Many issues are involved in the updating process, for example, how frequently to update, how to keep track of the historical versions, how to efficiently acquire the required data, how to automate the updating process, and how to disseminate updated data to end users. Most of these issues lie outside of the scope of this text except one, that is, the automation of the process. This topic will be further explored in Section 1.2.3.

1.2.3 TRANSFORMATIONS OF SPATIAL REPRESENTATIONS IN SCALE: GENERALIZATION

As discussed in Section 1.1.1, a spatial representation may take different forms, and a map is a typical type of spatial representation. Maps are associated with scale. Maps at different scales depict different levels of detail about the natural and cultural phenomena on the Earth. Figure 1.12 (See color insert following page 116) shows maps of Kowloon Peninsula of Hong Kong at two different scales. It can be seen clearly that the levels of abstraction are quite different in these two maps. Indeed, different symbols may be used to represent the same types of features but at different scale. This can be demonstrated by using the representation of a town as an example. It may be represented:

- By streets and building on maps at large scale.
- By main streets and big building blocks on maps at a smaller scale.
- By the outline of the town on maps at an even smaller scale.
- By a point symbol on maps at a small scale.
- By nothing as it disappears on maps at a very small scale.

This can be observed in Figure 1.9. Figure 1.9d is a large-scale representation, showing streets of Beijing in detail. Figure 1.9c is a map of the Beijing urban area,

(a) Topographic map 1:20,000 (HM20C)

(b) Topographic map 1:100,000 (HM100CL)

FIGURE 1.12 (See color insert following page 116) The Kowloon Peninsula represented on maps at two different scales (Courtesy of LIC of HKSAR).

showing major streets and city blocks. Figure 1.9b is map of Greater Beijing. In this representation, the urban area is simply outlined. Figure 1.9c is a map of China. In this representation, Beijing has almost become a point.

A national mapping agency may have maps at scales from 1:1,000 to 1:1,000,000 and even smaller. One critical issue faced by national mapping agencies is the frequent updating of maps at so many scales. The ideal approach is to update maps at the largest scale frequently and to derive maps at smaller scales on demand only. The process of deriving maps at a smaller scale from those at a larger scale is called *map generalization*.

All spatial representations are associated with scale. Therefore, generalization is a common issue for different types of spatial representation and is the process of multi-scale representation of spatial data. It has also been regarded by Li (1996) as the transformation of spatial representations in scale.

Generalization is a process of abstraction. The terrain features represented on maps at smaller scales are the abstractive representations of those at larger scales. Perhaps the highest level of generalization was achieved by Sir Isaac Newton, who represented any planet by a point in the derivation of his famous "Law of Gravitation." This means that generalization is an issue related not only to the representation of spatial data but also to the modeling of spatial processes.

In order to derive a spatial representation at a smaller scale from those at a larger scale, various types of transformations at different levels should be carried out. This also applies to the real-time zooming operation in spatial information systems. Therefore, in the later chapters, algorithms for such transformations will be presented.

1.3 TRANSFORMATIONS IN MULTI-SCALE SPATIAL REPRESENTATION

Before algorithms for different types of transformations can be presented, it is necessary to provide an overview of the operations required for the various transformations.

A spatial representation, for example, a map, contains the following types of information about features:

- (Geo)metric information related to position, size, and shape.
- Relational information about spatial relationships between neighboring features.
- Thematic information related to the types and importance of features.

Therefore, during the derivation of a spatial representation at a smaller scale from those at a larger scale, three types of transformations have been performed: geometric, relational, and thematic.

1.3.1 GEOMETRIC TRANSFORMATIONS

Geometric transformations are clearly demonstrated in Figures 1.9 and 1.12 and briefly discussed in Section 1.2.3. These transformations are achieved by some operations. The issues related to geometric transformations are:

- What operations are essential for multi-scale representation?
- What operations are currently available?
- What operations are required for a particular case?

The first question will be examined in detail in Section 1.4. The third one will be discussed in Section 1.3.3. The second question is discussed in this section.

In the classic textbook by Robinson et al. (1984), very few operations are identified, for example, classification, induction, simplification, and symbolization. However, these operations are too general to be computerized. That is, more concrete operations need to be identified. A summary of currently available operations is listed in Table 1.1. From this table, it can be seen clearly that some terms are in common use, while others are rarely used and that some terms are more specific, while others are more general.

Related to these operations of geometric transformations, a critical question is, "Is there a consensus on the use of these terms in the geospatial sciences (such as cartography and geographic information science)?" The answer is, "Not necessarily." This is revealed by a study carried out by Rieger and Coulson in 1993. In their study, a group of 23 expert cartographers from North America with various

TABLE 1.1
Operations for Geometric Transformations in Multi-Scale Representation (Su, B., Morphological Transformations for Generalization of Spatial Data in Raster Format. Ph.D. Thesis, Curtin University of Technology, Perth, Australia, 1997)

	Steward (1974)	Robinson et al (1984)	Delicia & Black (1987)	McMaster & Monmonior (1989)	Keates (1989)	Shea & McMaster (1989)	Beard & Mackaness (1991)	McMaster & Shea (1992)
Agglomeration			†					
Aggregation			†	†		†		†
Amalgamation				†		†		†
Classification	†	†				†	†	†
Coarsen					†		†	
Collapse			†	†	†	†	†	†
Combination					†		†	
Displacement				†		†	†	†
Enhancement				†		†		†
Exaggeration	†				†	†	†	†
Induction	†	†						
Merge				†		†		†
Omission				†	†		†	
Refinement			†	†		†		†
Selection	†				†		†	
Simplification	†	†	†	†	†	†		†
Smoothing				†		†		†
Symbolization	†	†						
Typification						†		†

backgrounds was interviewed regarding the use of the operations that frequently appear in literature: simplification, classification, displacement, selection, elimination, exaggeration, symbolization, smoothing, induction, and typification. Most of these terms were not defined in the same way by the experts, and they did not even understand a few of the terms. Indeed, some of the terms were defined in so many different ways that even the experts felt confused. For example, *simplification* has traditionally been used to mean the reduction of complexity (Keates, 1989) with the retention of the main structure. However, in recent years, it has been used to mean the reduction of points (McMaster and Shea, 1992). Therefore, there is a need for systematic classification and standardization. This will be addressed in Section 1.4.

For each operation, one or more algorithms may be developed. All algorithms for these operations together form a mathematical foundation for multi-scale spatial representation, leading to the title of this book. These algorithms can be stored as subroutines, such as the conformal and affine transformation models, and can be called whenever there is a need.

1.3.2 RELATIONAL TRANSFORMATIONS

After a geometric transformation is applied to a feature or a set of features, the relationship between neighboring features may have undergone a transformation (Dettori and Puppo, 1996). For example, if one generalizes buildings into street blocks, then the disjoint relation between individual buildings is changed, as the streets separating them disappear. There are three types of spatial relations:

- Topological relations: The connectivity and adjacency of spatial features.
- Order relations: The order between spatial features such as the directions, orientation, and comparison.
- Metric relations: The relations between metric properties of spatial features such as the distance and relative positions.

Spatial relations between features on a spatial representation may have changed after a geometric transformation. Inversely, such relations may be of help in the detection and resolution of spatial conflicts caused by geometric transformations.

Topological relations are the most fundamental (Freeman, 1975). Figure 1.13 shows an example of change in topological relations between area features. In the first instance, area features A and C were separated by feature B, but they became immediate neighbors after a transformation.

Eight basic types of topological relations between area features have been identified (Egenhofer and Franzosa, 1992). They are shown in Figure 1.14. Comparing Figure 1.13 with Figure 1.14, it can be seen that the change in topological relation in the case of Figure 1.13 is from "disjoint" to "meet."

Sometimes a change in topological relation may result in a *spatial conflict*. Figure 1.15 shows such an example. In this case, a small building falls into the water after geometric transformations. This can be detected by checking the topological relations between the water as an area feature and the small building as another area feature.

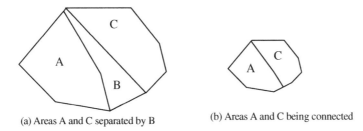

(a) Areas A and C separated by B   (b) Areas A and C being connected

FIGURE 1.13 Change in topological relation after a transformation in scale.

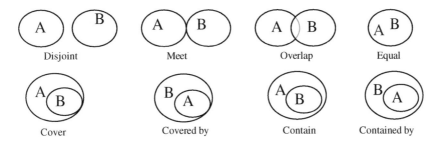

Disjoint   Meet   Overlap   Equal

Cover   Covered by   Contain   Contained by

FIGURE 1.14 Eight basic topological relations between area features A and B.

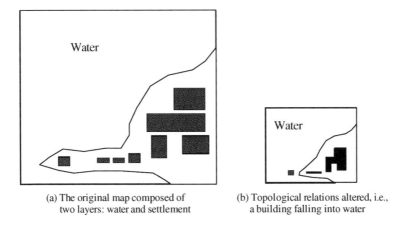

(a) The original map composed of   (b) Topological relations altered, i.e.,
two layers: water and settlement   a building falling into water

FIGURE 1.15 Spatial conflicts caused by a change in topological relation.

Comparing this with the cases shown in Figure 1.14, it can be seen that the topological relations between the water and the small building have changed from "disjoint" to "contain." This is geographically not acceptable in normal cases, and the problem needs to be solved. However, detailed discussion of the detection and resolution of spatial conflict lies outside the scope of this book.

To detect the change in topological relations, a mathematical model is required to describe the topological relations as shown in Figure 1.14. The classic model for formal description of topological relations is the *four-intersection model* by Egenhofer and Franzosa (1992). An extension to this model is the *nine-intersection model* (Egenhofer and Herring 1991). However, it has been pointed out (Chen et al., 2001) that the extension from four to nine intersections is invalid because there is linear dependency between the three topological components. It has also been pointed out that the four-intersection model cannot be used for all types of spatial features because the definitions of topological components are dimension dependent (Li et al., 2000). For example, in 1-D space, the two end points define the boundary of a line. However, this definition is not valid in 2-D space. If one simply adopts the definition in 1-D space to 2-D space, a topological paradox will be caused. To solve this problem, a *Voronoi-based spatial algebra* for topological relations has been developed by Li et al. (2002), which makes use of the features themselves and their *Voronoi regions* (see Section 2.3.3 for a more detailed discussion) only if necessary.

One interesting development is the *Voronoi-based k-order adjacency* model (Chen et al., 2004) for the more detailed differentiation of the disjoint relation. This model makes the use of Voronoi neighbors. In this model, the neighbors with one-order Voronoi-adjacency to a given feature are those features whose Voronoi regions are connected to these of the given feature. For example, the one-order Voronoi-adjacency neighbors of the given feature P in Figure 1.16a are the highlighted ones. The two-order Voronoi-adjacency neighbors of P are those features whose Voronoi regions are connected to the Voronoi regions of the one-order Voronoi-adjacency neighbors, as highlighted in Figure 1.16b. In another sense, this model might be regarded as a model for qualitative distance relations because it makes use of Voronoi regions of spatial features as a distance measure.

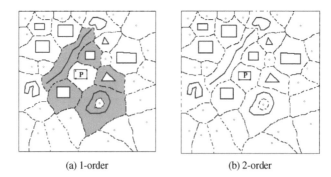

(a) 1-order   (b) 2-order

FIGURE 1.16 Voronoi-based k-order adjacency model to further differentiate the disjoint relation.

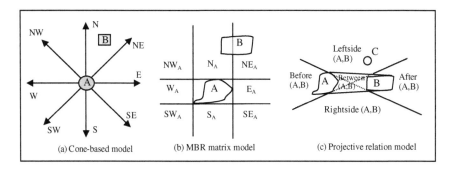

FIGURE 1.17 Directional relations between spatial features.

The model described above uses the term *order*. This is different from *order* in an ordinary sense, which means the arrangement of two or more features (or events) in accordance with stated criteria. For example, spatial features can be ordered according to size, distance, and direction. The *order relations* of spatial features can be described qualitatively or quantitatively. They can also be described in a relative sense or an absolute sense. Examples of relative descriptors are front and back, left and right, and above and below. The followings descriptors are absolute: N, NE, E, SE, S, SW, W, and NW.

Directional relations can be easily represented by a quantitative measure. The bearing in units of degree is a common practice. A direction is traditionally defined as a kind of distance-independent relation between two points in space that specifies the angular position of either with respect to the other. However, in a spatial representation, there are point, line, and area features. Attempts have been made to describe the directional relations among all these three types of spatial features. The cone-based four- and eight-direction models are in common use (e.g., Peuquet and Zhan, 1987). This is illustrated in Figure 1.17a, in which feature B is located on the north side of A if the four-direction model is used.

Sometimes the shape of an area is very complicated, and thus it is difficult to specify the directional relation in such a way. An alternative is the *MBR (minimum bounding rectangle) matrix model* (Goyal, 2000), as shown in Figure 1.17b (see Section 9.2.4 for a more detailed discussion). In this model an area is represented by its MBR. The whole space is divided into nine tiles. The relations are represented by a matrix. If feature B is located within the MBR of A, the directional relation is called "the same." In the case of Figure 1.17b, feature B is partially located in the N and NE directions. Weighting can also be used according to the proportion of the area in each tile.

A third model is the projective model used for reasoning the directional relations between a given area feature and other two area features (Billen and Clementini, 2004). It divides the space into five regions: before, after, left side, right side, and between. Figure 1.17c shows such a model, where feature C is on the left side of A and B. Another interesting development is the reduction of area features into line features by generalization for the establishment of directional relations between two area features (Yan et al., 2006).

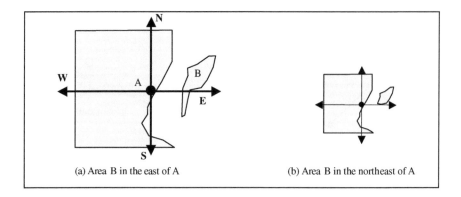

(a) Area B in the east of A   (b) Area B in the northeast of A

FIGURE 1.18 Change in directional relations after transformation in multi-scale representation.

Figure 1.18 shows the change in directional relations between areas A and B after a geometric transformation. With reference to the cone-based model, area feature B is to the east of A. After the geometric transformation in scale, B is completely on the northeast side of A.

Metric relations describe the relations between metric properties of spatial features such as distance and relative positions. The descriptors can be either quantitative or qualitative. The Voronoi-based k-order adjacency model is a kind of semiquantitative distance mode. Figure 1.19a shows another type of qualitative measure. However, it is based on the concept of distance.

The *Euclidean distance* is a quantitative metric between two points. Efforts have been made to extend the concept to describe the metric between lines and areas. Minimum distance, maximum distance, and centroid distance are widely used in geo-information science. It has also been suggested to make use of *Hausdorff distance* (Figure 1.19b) and *generalized Hausdorff distance* (Deng et al., 2005).

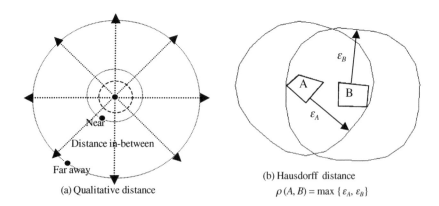

(a) Qualitative distance

(b) Hausdorff distance

$$\rho(A, B) = \max\{\varepsilon_A, \varepsilon_B\}$$

FIGURE 1.19 Distance relations between two spatial features.

It is clear that after a scale reduction the map space is reduced. The distance between two features is reduced, therefore, the representation needs to be modified. For example, some features are too close to be separated and thus need to be combined; some need to be displaced and others deleted. In other words, a set of operations for geometric transformation needs to be applied to make the result suitable for representation at a smaller scale. Such operations will be identified in Section 1.4.

1.3.3 THEMATIC TRANSFORMATIONS

After a geometric operation is applied to a feature or a set of features, the thematic meanings of features may have undergone a *thematic transformation*. For example, at a large scale, individual buildings are represented. Residential, commercial, and administrative buildings are identifiable. However, at a smaller scale, buildings of different types are grouped into city blocks. In this way, features with new thematic meanings are created and old features disappear. Similarly, different types of farm-lands are represented on land-use maps at large scales, such as irrigated land, irrigable land, and dry land. However, they may be aggregated into a new type, called farmland, at a smaller scale.

A reverse process can be applied to thematic information, that is, to make use of thematic information for formalization of rules to control geometric transforma-tions. For example, based on the biogeographical principle, there should be a piece of shrub land between lowlands and grassland (Pun-Cheng et al., 2003). This can be used as a rule for the transformations in spatial representation. Detailed discussion of thematic transformation lies outside the scope of this book, but more information can be found in Muller (1990), Buttenfield and McMaster (1991), Muller et al. (1995), Li and Choi (2002), and Gao et al. (2004).

1.4 OPERATIONS FOR GEOMETRIC TRANSFORMATIONS IN MULTI-SCALE SPATIAL REPRESENTATION

To derive small-scale spatial representations from large-scale ones, various types of geometric transformations need to be performed. A list of operations for geometric transformations is given in Table 1.1. However, as pointed out previously, no con-sensus has been made for some of these operations. Therefore, a systematic classi-fication and redefinition of essential operations needs to be carried out so that algorithms for these operations can be described in later chapters.

1.4.1 A STRATEGY FOR CLASSIFICATION OF OPERATIONS FOR GEOMETRIC TRANSFORMATIONS

Classification means to place together in categories those operations that resemble each other. A systematic classification should be complete (i.e., exhaustive),

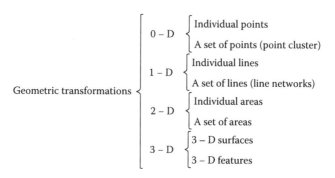

FIGURE 1.20 Classification of operations for geometric transformations in multi-scale representation based on the dimensions of geometric elements.

nonoverlapping, and objective. It is very difficult to have an exhaustive classification because no consensus has been made with the criteria for the assessment of multi-scale representation. Therefore, as one can imagine, the operations to be discussed in this section can only be regarded as essential operations.

Various criteria could be used for such a classification. For example, McMaster and Monmonior (1989) first used data mode as a criterion to classify these operations into two categories: raster-mode and vector-mode operations. Then they further identified a number of operations in raster and vector. However, the author believes that the data mode is not an issue anymore, and raster-based and vector-based algorithms can be easily integrated. Therefore, no special emphasis is needed on the difference in data mode. Instead, in this text emphasis is on the implementation of algorithms, and thus classification is determined based on the dimension of geometric elements, as shown in Figure 1.20.

1.4.2 Operations for Transformations of Point Features

For an individual point feature, the possible operations for geometric transformations are illustrated in Table 1.2 and their definitions are as follows:

- Displacement: To move a point away from a feature or features because the distance between the point and the other feature(s) is too small to be separated.
- Elimination: To eliminate an individual point feature, as it will be too small to represent.
- Magnification: To make the size of a point feature become large enough to be represented, although it appears too small to be represented.

In some literature magnification is also called exaggeration or enlargement.

For a set of point features, the operations are illustrated in Table 1.3 and the definitions are as follows:

TABLE 1.2
Operations for Geometric Transformations of Individual Point Features

Operators	Large-scale	Photo-reduced	Small-scale
Displacement			
Elimination			
Magnification			

TABLE 1.3
Operations for Geometric Transformations of a Set of Point Features

Operation	Large-scale	Photo-reduced	Small-scale
Aggregation			
Regionalization			
Selective omission			
(Structural) Simplification			
Typification			

- Aggregation: Grouping a number of points into a single point features.
- Regionalization: Outlining the boundary of the region covered by points so as to make this region an area feature.
- Selective omission: Selection of more important point features to be retained and omission of less important ones, if space is not adequate.
- Simplification: Reducing the complexity of the structure of a point cluster by removing some point, with the original structure retained.
- Typification: Keeping the typical pattern of the point feature while removing some points.

1.4.3 OPERATIONS FOR TRANSFORMATIONS OF LINE FEATURES

Representation of lines in multi-scale has been a heavily researched topic for a long time because (a) it is a staring point for new studies of the multi-scale representation

TABLE 1.4

Operations for Geometric Transformations of Individual Line Features

Operation		Large-scale	Photo-reduced	Small-scale
Displacement				
Elimination				
(Scale-driven) generalisation				
Partial modification				
Point reduction				
Smoothing	Curve fitting			
	Filtering			
Typification				

of other features and (b) over 80% of features on a map are line features. A huge body of literature is available on the manipulation of line features.

For an individual line feature, the possible operations for geometric transformation are illustrated in Table 1.4, and the definitions are as follows:

- Displacement: Moving the line in a given direction.
- Elimination: Eliminating an individual line feature because it will be too narrow to represent.
- (Scale-driven) generalization: Producing a new line in which the main structure is retained but small details are removed. This operation is dependent on the scales of input and output representations.
- Partial modification: Modifying the shape of a segment within a line.
- Point reduction: Reducing the number of points for representation by removing the less important points from a line so that only the so-called critical points are retained.
- Smoothing: Making the line appear smoother.
- Typification: Keeping the typical pattern of line bends while removing some.

There are two types of smoothing: *filtering* and *curve fitting*. *Filtering* means to filter out the high-frequency component (or small details) of a line so that the line appears smoother. *Curve fitting* is another type of smoothing, which tries to fit a curve through a set of points. In the author's viewpoint, smoothing is not necessarily an operation for multi-scale representation of lines. However, smoothing does create some effects required for multi-scale representation, although this operation is not directly related to scale.

The term *point reduction* is not in wide use. In some literature, *line simplification* is used to refer to point reduction, as some kind of simplification might be created in some cases. In computing literature, curve approximation and corner detection are the two operations used to retain critical points and remove less important points. In the author's viewpoint, point reduction should not be part of the operations for multi-scale representation because traditional generalization has nothing to do with point reduction (Li, 1993). Point-reduction algorithms try to make best approximations of the original line with a minimum number of points. It must be emphasized here that no scale change is involved in such an operation. The output is for the representation of the line at the same scale. These algorithms are good for *weeding* operations. Reduction of the number of points on a line was an important issue in the early development of digital representation because of limited computing power. Indeed, at that time, data volume was a big concern. As a consequence, many algorithms have been developed for such a purpose. In practice, point reduction should also be applied to lines to reduce the number of points as preprocessing of scale-driven generalization because the points on a line will appear to be too dense after a scale reduction, as can be seen from Table 1.4.

Scale-driven generalization is a type of smoothing. In scale-driven generalization, the smoothing effect is computed based on the scale of the input data and the scale of the output data. In the end of such a process, the main trend of the line is retained and small variations removed.

For a set of line features, possible operations for geometric transformations are illustrated in Table 1.5, and the definitions are as follows:

- Selective omission: Selecting the more important lines to be retained.
- Collapse:Making the dimension changed. Two types are identified: ring-to-point and double-to-single-line.
- Enhancement: Making the characteristics still clear.
- Merging: Combine two or more close lines together.
- Displacement: Moving one away from the other or both lines away from each other.

In Table 1.5 the example of selective omission is a river network. There is also a selective omission problem for contour lines and a transportation network.

1.4.4 OPERATIONS FOR TRANSFORMATIONS OF AREA FEATURES

The operations for individual area features are listed in Table 1.6 and are defined as follows:

- Collapse: Making the feature represented by a symbol with lower dimension.
- Displacement: Moving the area to a slightly different position, normally to solve the conflict problem.

TABLE 1.5
Operations for Geometric Transformations of a Set of Line Features

Operation		Large-scale	Photo-reduced	Small-scale
Selective omission				
Collapse	Ring-to-point			
	Double-to-single			
Enhancement				
Merging				
Displacement				

TABLE 1.6
Operations for Geometric Transformations of Individual Area Features

Operation		Large-scale	Photo-reduced	Small-scale
Collapse	Area-to-point			
	Area-to-line			
	Partial			
Displacement				
Exaggeration	Directional thickening			
	Enlargement			
	Widening			
Elimination				
(Shape) Simplification				
Split				

- Exaggeration: Making an area with small size still represented at a smaller scale on maps on which it should be too small to be represented.
- Elimination: Eliminating small and unimportant areas.
- Simplification: Making the shape simpler.
- Split: Splitting an area's features into two because the connection between them is too narrow.

There are three types of collapse: *area-to-point collapse* (e.g., representing a city with a point feature on small-scale maps), *area-to-line collapse* (e.g., representing a river with a single line), and *partial collapse* (representing the thin part of an area feature with a line while the other part is still a region).

There are also three types of exaggeration: *directional thickening* (making the area feature exaggerated in a given direction), *enlargement,* or magnification (making the whole feature enlarged in all directions), and *widening* (making the bottleneck of an area feature wider to make it observable at a smaller scale).

For a set of area features, in addition to simplification, displacement, selective omission, collapse, and exaggeration, the following form a subset of operations:

- Agglomeration: Making area features bounded by thin area features into adjacent area features by collapsing the thin area boundaries into lines.
- Aggregation: Combining area features (e.g., buildings) separated by open space.
- Amalgamation: Combining area features (e.g., buildings) separated by another feature (e.g., a road).
- Dissolving: Splitting a small area into pieces (according to adjacent areas) and merging these pieces into their corresponding adjacent areas.
- Merging: Combining two adjacent areas into one.
- Relocation: Moving more than one feature around normally to solve a conflict problem.
- Structural simplification: Retaining the structure of area patches by selecting important ones and omitting less important ones.
- Typification: Retaining the typical pattern, for example, a group of area features (e.g., buildings) aligned in rows and columns.

1.4.5 OPERATIONS FOR TRANSFORMATIONS OF 3-D SURFACES AND FEATURES

For 3-D surfaces, two types of multi-scale representation are differentiated (Li et al., 2005): metric and visual. In *metric multi-scale representation,* the features on the same representation have the same scale. Filtering and pyramid structuring are the methods commonly used (de Floriani, 1989). Scale-driven generalization has also been discussed by Li and Li (1999) and Li et al. (2005). However, in *visual multi-scale representation* the features on the same representation may have different scales. In other words, the scale of the representation may vary from place to place on the same representation. *Level of detail* (LoD) is the concept used to refer to such visual multi-scale representations (Luebke et al., 2003).

TABLE 1.7
Operations for Geometric Transformations of a Set of Area Features

Operation	Large-scale	Photo-reduced	Small-scale
Aggregation			
Agglomeration			
Amalgamation			
Dissolving			
Merging			
Relocation			
(Structural) Simplification			
Typification			

For 3-D features, the operations identified by researchers are still similar to those for area features listed in Table 1.7. However, some terms may have slightly different meanings. For example, *exaggeration* has been used to refer to enlarging the size of the doors of a building instead of the building itself (Bai and Chen, 2001). Such exaggeration will be defined as *partial exaggeration* in this context. In addition, *bunching* and *injoining* have been in use (Bai and Chen, 2001), but they are similar to the *typification* and *aggregation* operations for 2-D representations.

In essence, there is not much difference between the operations used for 2-D and 3-D representations. Therefore, no further discussion of the operations for the geometric transformation of 3-D features will be conducted here.

1.5 SCOPE OF THIS BOOK

In Section 1.4 essential operations for geometric transformations of spatial representation are systematically classified. One or more algorithms are required for each of these operations. This book is an attempt to provide a comprehensive coverage of these algorithms. Only the *low-level algorithms* (or operators) for such transformations will be presented. *High-level algorithms*, that is, those based on neural networks and compound algorithms, are not included. High-level rules for controlling the algorithms and the spatial relations between the spatial features are not also discussed.

In order to make the description of such algorithms more convenient, mathematical tools that are widely used for algorithm development are included in Chapter 2, and some principles and strategies are presented in Chapter 3. From Chapter 4 on, algorithms for multi-scale spatial representations will be presented. The presentations are organized according to the classifications described in Section 1.4.

Chapter 4 is dedicated to the multi-scale representation of point features. The elimination of individual point features is an easy operation and there is no need of any algorithm. The displacement of a point feature is similar to displacement of a line or an area feature and will be discussed in Chapter 11, which is dedicated to the topic of displacement. The magnification of a point feature means the enlargement of a small area feature and will be discussed in Chapter 9. Therefore, in that chapter, only algorithms for a set of point features are presented. Typification will be discussed in Chapter 10, which is dedicated to a set of area features because point features under typification are small area features.

Lines have been well studied because they are the most frequently occuring features on a topographic map and they can also be used to represent area features (i.e., by boundaries). Various types of treatments have been made to lines. In this text, three chapters are dedicated to individual lines. Chapter 5 presents some algorithms for point reduction, Chapter 6 for line smoothing, and Chapter 7 for scale-driven generalization. Displacement and partial modification are discussed in Chapter 11. Typification of line bends is omitted here because of its subjectivity. Readers who are interested in this operation are referred to the articles by Plazanet et al. (1995) and Burghardt (2005). Chapter 8 presents algorithms for multi-scale representation of three types of line networks: contours, hydrological networks (i.e., rivers), and transportation networks (i.e., roads).

Chapter 9 presents algorithms for the multi-scale representation of individual area features and Chapter 10 for the multi-scale representation of area features at the class level.

Chapter 11 is dedicated to displacement. Algorithms for various types of displacement are presented that are common for point, line, and area features. The last chapter, Chapter 12, presents some more recent developments for 3-D surfaces and 3-D features.

REFERENCES

Abler, R., The National Science Foundation National Center for Geographc Information and Analysis, *International Journal of Geographical Information Systems,* 1(4), 303–326, 1987.

Bai, F. W. and Chen, X. Y., Generalization for 3D GIS, in *Proceedings of the 3rd ISPRS Workshop on Dynamic and Multi-dimensional GIS & the 10th Annual Conference of CPGIS on Geoinformatics*, 23–25 May 2001, Bangkok, Thailand, Chen, J., Chen, X. Y., Tao, C., and Zhou, Q. M., Eds., 2001, pp. 8–11.

Beard, M. K. and Mackaness, W., Generalization operators and supporting structures, *Proceedings Auto-Carto* 10, 29–45, Baltimore, MD, 1991.

Billen, R. and Clementini, E., Introducing a reasoning system based on ternary projective relations, in *Developments in Spatial Data Handling*, Fisher, P., Ed., Springer, Berlin, 2004, pp. 381–394.

Burghardt, D., Controlled line smoothing by snakes, *GeoInformatica*, 9(3), 237–252, 2005.

Burrough, P. A. and McDonnell, R., *Principles of Geographical Information Systems*, 2nd ed., Oxford Press, Oxford, UK, 1998.

Buttenfield, B. P. and McMaster R. B., *Map Generalization: Making Rules for Knowledge Representation*. Longman Scientific and Technical, London, 1991.

Chen, J., Li, C. M., Li, Z. L., and Gold, C., A Voronoi-based 9-intersection model for spatial relations, *International Journal of Geographical Information Science*, 15(3), 201–220, 2001.

Chen, J., Zhao, R. L. and Li, Z. L., Voronoi-based K-order neighbour relations for spatial analysis, *ISPRS Journal of Photogrammetry and Remote Sensing*, 59(1-2), 60–72, 2004.

Christensen, A. H., Line generalization by waterline and medial-axis transformation: success and issues in an implementation of Perkel's proposal, *The Cartographic Journal*, 26(1), 19–32, 2000.

de Floriani, L., A pyramidal data structure for triangle-based surface description, *IEEE Computer Graphics and Applications*, 9(2), 67–78, 1989.

Deng, M., Chen, X. Y., and Li, Z. L., A generalized Hausdorff distance for spatial objects in GIS, in *International Archives of Photogrammetry and Remote Sensing, Vol. XXXVI, Part 2/W29, (Proceedings of the 4th Workshop on Dynamic and Multi-dimensional GIS)*, Pontypridd, UK, 2005,10–15.

Dettori, G. and Puppo, E., How generalization interacts with the topological and metric structure of maps, in *Proceedings of SDH'96, 12-16 August 1996*, The Netherlands, 1996, pp. 9A.27–9A.38.

DiBiase, D. W., MacEachren, A. M., Krygier, J. B., and Reeves, C., Animation and the role of map design in scientific visualisation, *Cartography and Geographic Information Systems*, 19, 201–214, 265–266, 1992.

Egenhofer, M. and Franzosa, R., Point-set topological spatial relations, *International Journal of Geographical Information Systems*, 5(2), 161–174, 1992.

Egenhofer, M. and Herring, J., Categorizing Binary Topological Relationships between Regions, Lines, and Points in Geographic Databases, technical report, Department of Surveying Engineering, University of Maine, Orono, 1991.

Frank, A. U., Qualitative spatial reasoning about distances and directions in geographic space, *Journal of Visual Languages and Computing*, 3(2): 343-371, 1992.

Freeman, J., The modelling of spatial relations, *Computer Graphics and Image Processing*, 4, 156–171, 1975.

Gao, W. X, Gong, J. Y. and Li, Z. L, 2004. Thematic knowledge for the generalization of land-use map, *Cartographic Journal*, 41(3), 245–252, 2004.

Goyal, R., Similarity Assessment for Cardinal Directions between Extended Spatial Objects, PhD Thesis, University of Maine, Orono, 2000.

Keates, J., *Cartographic Design and Production*, 2nd ed., Longman Scientific, Harlow, UK, 1989.

Kraak, M.-J., Settings and needs for web display, in *Web Cartography: Developments and Prospects,* Kraak, M.-J. and Brown, A., Eds., Taylor & Francis, London, 2001, pp. 1–7.

Lee, Y. C., Kwong, A., Pun, L., and Mack, A., Multi-media map for visual navigation, *Journal of Geospatial Engineering*, 3(2), 87–96, 2001.

Li, Z. L., Some observations on the issue of line generalisation, *Cartographic Journal*, 30(1), 68–71, 1993.

Li, Z. L., Reality in time-scale system and cartographic representation, *Cartographic Journal*, 31(1), 50–51, 1993.

Li, Z. L., Transformation of spatial representation in scale dimension: a new paradigm for digital generalization of spatial data, *International Archives for Photogrammetry and Remote Sensing*, XXXI(B3), 453–458, 1996.

Li, Z. L. and Choi, Y. H., Topographic map generalisation: association of road elimination with thematic attributes, *Cartographic Journal*, 39(2), 153–166, 2002.

Li, Z. L. and Li, C. M., Objective generalization of DEM based on a natural principle, in *Proceedings of 2nd International Workshop on Dymanic and Multi-dimensional GIS*, Oct. 4–6, 1999, Beijing, 1999, pp.17–22.

Li, Z. L., Li, Y. L., and Chen, Y. Q., Basic topological models for spatial entities in 3-dimensional space, *GeoInformatica*, 4(4), 419–433, 2000.

Li, Z. L., Zhao, R. L., and Chen, J., A Voronoi-based spatial algebra for spatial relations, *Progress in Natural Science*, 12(7), 528–536, 2002.

Li, Z. L., Zhu, Q., and Gold, C., *Digital Terrain Modelling: Principles and Methodology*, CRC Press, Boca Raton, FL, 2005.

Luebke, D., Reddy, M., Cohen, J., Varshney, A., Watson, B., and Huebner, R., *Level of Detail for 3D Graphics*, Morgan Kaufmann, San Francisco, 2003.

McMaster, R. and Monmonior, M., A conceptual framework for quantitative and qualitative raster-mode generalisation, in *Proceedings of GIS/LIS'89*, Orlando, Florida, 1989, pp. 390–403.

McMaster, R. B. and Shea, K. S., *Generalization in Digital Cartography*, Association of American Geographers, Washington, DC, 1992.

Moellering, H., Real maps, virtual maps, and interactive cartography, in *Spatial Statistics and Models*, Gaile, G., Ed., Kluwer Academic, Dordrecht, The Netherlands, 1984, pp. 109–131.

Moellering, H., Understanding modern cartography using the concepts of real and virtual maps, in *Proceedings of the XIII International Cartographic Conference*, Morelia, Mexico, 1987, pp. 43–52.

Muller, J. C., Rule based generalization: potentials and impediments, in *Proceedings of 4th International Symposium on Spatial Data Handling, Zurich, July 22–23* 1990, pp. 317–334.

Muller J. C., Lagrange J. P., and Weibel, R., *GIS and Generalization: Methodology and Practice*, Taylor & Francis, Bristol, UK, 1995.

Peuquet, D. and Zhan, C. X., An algorithm to determine the directional relation between arbitrarily-shaped polygons in the plane, *Pattern Recognition*, 20(1), 65–74, 1987.

Plazanet, C., Affholder, J. G., and Fritsch, E., The importance of geometric modelling in linear feature gneralization, *Cartography and Geographic Information Systems*, 22(4), 291–305, 1995.

Pun-Cheng, L., Li, Z. L., and Gao, W., Integration of generalization operators for vegetation maps based on bio-geographical principles, *Cartographica*, 32(2), 17–30, 2003.

Rieger, M. and Coulson, M., Consensus or confusion: cartographers' knowledge of generalisation, *Cartographica*, 30(2), 69–80, 1993.

Robinson, A. H., Sale, R., Morrison, J. L., and Muehrcke, P. C., *Elements of Cartography*, 5th ed., Wiley, New York, 1984.

Shea, K. S. and McMaster R. B., Cartographic generalisation in a digital environment: when and how to generalise, in *Proceedings of Auto Carto 9*, Baltimore, MD, 1989, pp. 56–67.

Steward, H. J., Cartographic generalization: some concepts and explanations, *Canadian Cartographer*, 11(10) (Suppl. 1), 1974.

Su, B., Morphological Transformations for Generalization of Spatial Data in Raster Format. PhD thesis, Curtin University of Technology, Perth, Australia, 1997.

van den Worm, J., Web map design in practice, in *Web Cartography: Developments and Prospects*, Kraak, M. J. and Brown, A., Eds., Taylor & Francis, London, 2001, pp. 87–107.

Yan, H., Chu, Y., Li, Z. L., and Guo, R., A quantitative description model for direction relations based on direction groups, *GeoInformatica*, 10(2), 177–195, 2006.

2 Mathematical Background

In the algorithms to be presented in the later chapters, mathematical tools at various levels are involved. To facilitate those discussions, this chapter provides some basic mathematical background.

2.1 GEOMETRIC ELEMENTS AND PARAMETERS FOR SPATIAL REPRESENTATION

2.1.1 COORDINATE SYSTEMS

To make a spatial representation possess a certain level of metric quality, a *coordinate system* needs to be employed. The *Cartesian coordinate system* is the basic system in Euclidean space and is most familiar to us. It could be three-dimensional (3-D) or 2-D. The latter is a result of the orthogonal projection of the former. A *geographical coordinate system* is also a fundamental system for spatial representation, consisting of longitude and latitude. A geographical coordinate system can be defined on a sphere (or spheroid) or on a 2-D plane. The latter is a result of a projection of the former. Such a projection is called a *map projection*. *Polar coordinate systems* are also possible but are not widely used in spatial representation. Figure 2.1 shows such systems in a 2-D plane. The Cartesian coordinate system is normally used for spatial representation at large and medium scales, and the geographical coordinate system is used for spatial representation at small and very small scales.

In recent years *raster systems* have been in wide use. Through the history of spatial information science, there has been an almost constant debate about the nature and importance of raster (e.g., Peuquet, 1984; Maffini, 1987; Maguire et. al., 1991). The importance of working with multi-scale representation in raster was realized in the early 1980s (e.g., Monmonior, 1983). Mathematically elegant algorithms have been developed since the 1990s (Li, 1994; Li and Su, 1996; Su et al., 1997a, 1997b, 1998).

A *raster space* is a discrete space. It is a result of partitioning a connected space into small pieces, but these pieces together cover the whole space contiguously. This kind of partition is called *space tessellation*. Various approaches are possible for space tessellation (Lee and Li, 1998), such as grids, hexagons, triangular irregular networks, and Voronoi diagrams, as shown in Figure 2.2. (The latter two will be discussed in Section 2.3). Raster, as a grid-based tessellation, is only one among many possible alternatives.

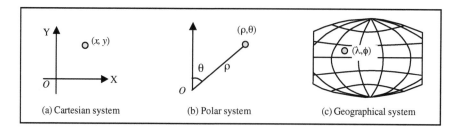

FIGURE 2.1 Coordinate systems used in 2-D spatial representation.

The tessellation based on squared grids constitutes raster space. Each grid is viewed as a *spatial element* (abbreviated as *spel*) or *picture element* (abbreviated as *pixel*) in the geoinformation community.

In 2-D space the adjacency between two neighboring pixels could be defined in two different ways, that is, four-adjacency or eight-adjacency (see Figure 2.3). With four-adjacency, a pixel is considered as being connected to neighboring pixels along only four sides: top, bottom, left, and right (Figure 2.3a). However, with eight-adjacency, a pixel is considered as being connected to neighboring pixels along four diagonal directions: upper left, upper right, lower right, and lower left (Figure 2.3b). In 3-D space there are 18-adjacency and 26-adjacency. A more detailed discussion regarding this issue can be found in the paper by Kong and Rosenfeld (1989).

2.1.2 Representation of Geometric Elements in Vector and Raster Spaces

On a spatial representation spatial features are represented by basic *geometric elements*, that is, points, lines, and areas. A point is used to indicate the position of a spatial feature, a line to indicate a position and a direction, and an area to indicate a position, direction, and extent.

A spatial feature can be presented in vector or raster space. In vector space a point is represented by two coordinates, that is, *P(x,y)*, and has infinitesimal size. A straight line is determined by two points, and a plane by three points. A curved line

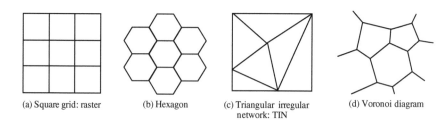

FIGURE 2.2 Raster as a particular type of space tessellation.

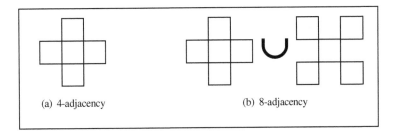

FIGURE 2.3 Connectedness between raster pixels. (a) Four-adjacency and (b) eight-adjacency.

may consist of a set of connected straight line segments, but it has no thickness. An area is a part of a plane and is represented by its boundary—a closed line. Figures 2.4a–c shows the representations of two points, a line, and an area in vector space, respectively.

In raster space a point is represented by a pixel with certain size, and the coordinates of a point are defined by two integer numbers, that is, row and column. A line is represented by a string of pixels, and an area by a region with a group of pixels. Such representations are illustrated in Figures 2.4d-f

2.1.3 SOME COMMONLY USED GEOMETRIC PARAMETERS

The mathematical functions of the basic geometric elements used for representation of spatial features are listed in Table 2.1, in which a point is represented by $P(x,y)$. In row 1 x_0, y_0 is the center of the circle and r is the radius.

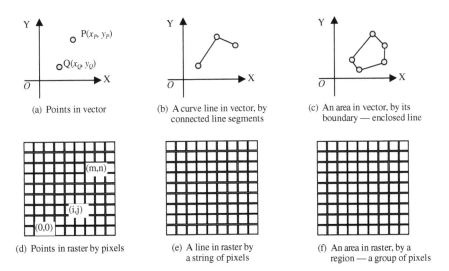

FIGURE 2.4 Representation of spatial features in vector and raster spaces.

TABLE 2.1

Mathematical Functions for Geometric Elements

	Geometric elements	Mathematical function
1	Straight line (*Ln*)	$ax + by + c = 0$
2	Plane (*Pl*)	$ax + by + cz + d = 0$
3	Curved line (*Cl*)	$y = a + bx + cx^2 + ...$
4	Circle (*Cr*)	$(x - x_0)^2 + (y - y_0)^2 = r^2$
5	Curved surface (*Cs*)	$z = a_0 + a_1 x + b_1 y + a_2 x^2 + b_2 y^2 + ...$

A line in a vector can also be represented in a parametric form as follows:

$$g(L) = \begin{cases} X(L) = f_x(L) \\ Y(L) = f_y(L) \end{cases} \tag{2.1}$$

where L is the length along the curved line, sometimes called the *arc length*. In this equation each of the X and Y coordinates has become a *monotonic function* of L.

Figure 2.5 shows the *parametric representation* of a curved line. Figure 2.5a is a curved line formed by a set of points. Such a line is sometimes called a *polyline*. Figure 2.5b shows the decomposition of the polyline line into two monotonic functions.

In any algorithm for geometric transformations, some kinds of geometric parameters must be used as criteria. Algorithms for multi-scale spatial representation are

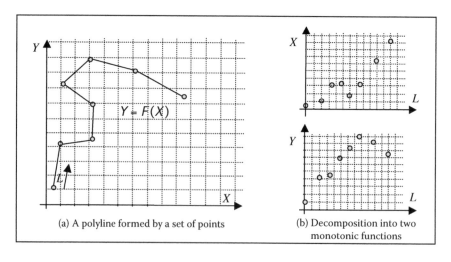

(a) A polyline formed by a set of points

(b) Decomposition into two monotonic functions

FIGURE 2.5 Representation of a line in parametric form.

TABLE 2.2
Some Commonly Used Geometric Parameters

	Geometric parameter	Mathematical function
1	Distance between two points	$d(P_1,P_2) = \sqrt{(x_1 - x_2)^2 + (y_1 - y_2)^2}$
2	Distance from point P to line Ln	$d(P,Ln) = \dfrac{\|ax_1 + by_1 + c\|}{\sqrt{a^2 + b^2}}$
3	Distance from point P to plane Pl	$d(P,Pl) = \dfrac{\|ax_1 + by_1 + cz + d\|}{\sqrt{a^2 + b^2 + c^2}}$
4	Slope between two points	$\tan\alpha = \dfrac{y_2 - y_1}{x_2 - x_1}$
5	Curvature of a curved line	$c(x,y) = \dfrac{d^2 y/dx^2}{[1 + (dy/dx)^2]^{3/2}}$
6	Angle (ω) formed by two sides (a and b) of a triangle	$\cos\omega = \dfrac{a^2 + b^2 - c^2}{2ab}$
7	Area formed by N points	$A(P_1,P_2\cdots P_N) = \dfrac{1}{2}\sum_{i=1}^{N}\left(y_i \times x_{i+1} - x_i \times y_{i+1}\right)$

not exceptions. Table 2.2 lists some parameters commonly used in algorithm development. In this table, a point is represented by $P(x,y)$; the term *distance* in rows 1–3 means the shortest Euclidean distance; the α in row 4 is the slope angle; the ω in row 6 is the angle opposite the side c. In row 7 the N points that form an area should be arranged in a clockwise direction, and the $(N + 1)^{th}$ point means the first point.

In raster space the distance between two points at $P_1(i,j)$ and $P_2(m,n)$ can also be computed using the Euclidean distance function as shown in row 1 of Table 2.2:

$$d(P_1,P_2) = f(i,j,m,n) = \sqrt{(i - m)^2 + (j - n)^2} \qquad (2.2)$$

The unit of $d(P_1,P_2)$ is a pixel. For example, if the two points are at (2,2) and (3,3), then the result is $\sqrt{2}(\approx 1.414)$ pixels. This result in decimal form is inconvenient to use in raster; a distance in integer numbers is more desirable and thus normally employed. A number of definitions for raster distance have been proposed (Rosenfeld and Pfaltz, 1968; Borgefors, 1986, 1994; Melter, 1987; Breu et al., 1995; Embrechts and Roose, 1996), as shown in Figure 2.6.

(a) Chessboard   (b) City-Block   (c) Octagon   (d) Chamfer 2-3   (e) Chamfer 3-4

FIGURE 2.6 Various definitions of raster distances.

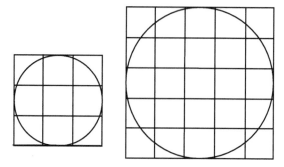

FIGURE 2.7 Difference between Euclidean distances and chessboard distances.

The concept of *raster distance* is directly related to the number of neighbors or the directions of connection. Suppose that one is traveling from the point at (2,2) to the point at (3,3); there is only one step if one is allowed to travel along the diagonals. In this case the most appropriate raster distance between these two points is 1 (pixel). This is the case with eight-ajacency. A diagrammatic representation of this type of distance is given Figure 2.6a. The shape of this diagram is like a chessboard—thus the name, *chessboard distance*. However, in the case four-ajacency, one is only allowed to travel along four directions (i.e., not along diagonals), so from (2,2), one needs to travel via either point (2,3) or (3,2) to point (3,3). Two steps are involved, so the most appropriate distance between (2,2) and (3,3) is 2 in this case. The resulting distance diagram is shown in Figure 2.6b. This is what happens when one is traveling in a city. That is, one has to travel along streets — thus the name, *city block distance*.

It is obvious that the approximation of these two types of distances to the Euclidean distance becomes poorer and poorer when the distance becomes larger and larger. Figure 2.7 illustrates the difference between the chessboard distance and Euclidean distance. To make such an approximation more accurate, other distance definitions have been proposed such as Chamfer 2-3, Chamfer 3-4, octagonal distance, and so on (see Borgefors, 1986), as shown in Figures 2.6d and 2.6e.

2.1.4 DIMENSIONALITY OF SPATIAL FEATURES

In Euclidean geometry spatial features are represented by basic geometric elements, that is, point, line, area, and volume, with *dimensionality* of 0, 1, 2, and 3, respectively. The question that comes up is, "Why is this?"

To answer this question, one needs to discuss the meanings of *dimension* in the first place. One could understand the dimension as the linearly independent directions to travel (move). For example, one plane has a two linearly independent directions and thus has a dimension of two.

Alternatively, one could think in terms of self-similarity. Suppose one divides a line segment into four self-similar pieces, each with the same length. One could obtain the original line segment from any one of the four pieces if it is magnified by four times.

Similarly, if one divides a line segment into 16 self-similar pieces, each with the same length, one could obtain the original line segment from any of the 16 pieces with a 16-times magnification. In the case of a plane, the situation is different. Suppose one divides a square into 16 self-similar subsquares; one needs to use only a four-times magnification to obtain the original square from any of these 16 subsquares. From these examples, one could define dimension as follows:

$$D = \frac{\log(number\ of\ self-similar\ pieces)}{\log(magnification\ factor)} = \frac{\log n}{\log m} \tag{2.3}$$

For the case of line segment, $n = 4$, $m = 4$, $D_{Ln} = \log4/\log4 = 1$ in the first case, and $n = 16$, $m = 16$, $D_{Ln} = \log16/\log4 = 1$ in the second case. However, in the case of a square, $n = 16$, $m = 4$, $D_{Pl} = \log16/\log4 = \log4^2/\log4 = 2$. Similarly, a value of 3 could be obtained for the dimension of a volume.

However, a line in geographical space is more complicated than a straight line. Richardson (1961) found that the measured lengths (L) of a coastline (such as those of Australia, Britain, Germany, Portugal, and South Africa) are dependent on the step size (δ) of the map divider that is used to step along the line (on a map) for measurement and found the empirical formula as follows:

$$L \propto \delta^{-\alpha} \tag{2.4}$$

where, α is a constant. Through the interpretation of this constant, Mandelbrot (1967, 1977, 1983) developed a new concept called fractal dimension and a new branch of geometry called fractal geometry. In this new geometry the value of a dimension can be a fractal number. A curved line has a dimension from 1 to 2, and a curved surface has a dimension between 2 and 3, depending on the complexity. Figure 2.8 shows the dimensionality of the spatial features. Figure 2.9 shows the Koch line

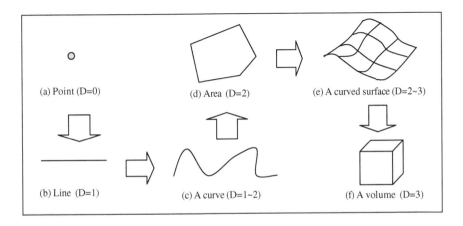

FIGURE 2.8 Dimensionality of spatial features.

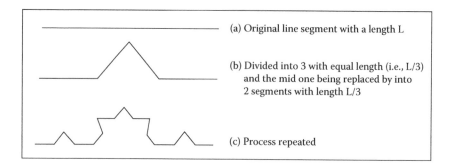

FIGURE 2.9 A curved line with 1.26 dimension (Koch curved line).

with a dimension of 1.26. That is, $n = 4$, $m = 3$, $D_{Ln} = \log4/\log3 = 1.26$. The value of D has been used to measure the complexity of curved lines and curved surfaces (e.g., Muller, 1986; Goodchild, 1980; Buttenfield, 1989; Carstensen, 1989).

Mandelbrot (1967, 1977, 1983) presented an alternative expression for the relationship between the measured lengths (L) of the coastlines and step size (δ) of the map divider as follows:

$$L \propto \delta^{1-D} \tag{2.5}$$

where D is a fractal dimension of the line. Equation 2.5 can also be rewritten as

$$L = K\delta^{1-D} \tag{2.6}$$

where K is a constant.

In practice, one important issue is how to compute or estimate the fractal dimension of a line feature. If a line segment can be divided into N identical parts, each of which is scaled down by a factor $\delta = 1/N$ from the whole, then the fractal dimension (D) of the line is

$$D = \frac{\log N}{\log(1/\delta)} \tag{2.7}$$

Suppose two map dividers with step sizes of δ_1 and δ_2, respectively, are used to measure a coastline. The numbers of steps for the corresponding measurements are n_1 and n_2. Then the fractal dimension of the coastline can be computed as follows (Goodchild and Mark, 1987):

$$D = \frac{\log(n_2/n_1)}{\log(\delta_1/\delta_2)} \tag{2.8}$$

For example, in Figure 2.9b the step size of map divider $\delta_b = L/3$, and the resultant number of steps $n_b = 4$. In Figure 2.9c the step size of map divider $\delta_c = (L/3)/3 = L/9$, and the resultant number of steps $n_c = 16$. The fractal dimension D of this line is $\log(16/4)/\log[(L/3)/(L/9)] = \log4/\log3 = 1.26$.

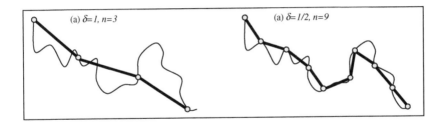

FIGURE 2.10 A curved line with fractal dimension 1.58.

Figure 2.10 is another example showing the computation of fractal dimension. In Figure 2.10a $\delta = 1$ and $n = 3$. In Figure 2.10b $\delta = 1/2$ and $n = 9$. The fractal dimension of this curve line is $D = \log(9/3)/\log(1 \div 1/2) = \log3/\log2 = 1.58$.

Suppose a number of map dividers with step size $\delta_1, \delta_2, \delta_3, \ldots,$ and δ_M are used to measure a line; the corresponding measured lengths are $L_1, L_2, L_3, \ldots, L_M$. Then by applying a regression operation to Equation 2.6, a formula to compute the fractal dimension can be expressed as follows (Wang and Wu, 1998):

$$D = 1 - \frac{\sum_{i=1}^{M} [\log(\delta_i)\log(L_i)] - \frac{1}{M} \times \left(\sum_{i=1}^{M} \log(\delta_i)\right) \times \left(\sum_{i=1}^{M} \log(L_i)\right)}{\left(\sum_{i=1}^{M} \log(L_i)\right)^2 - \frac{1}{M} \times \left(\sum_{i=1}^{M} \log(\delta_i)\right)^2} \qquad (2.9)$$

where M is the number of map dividers used for measurement.

Similar to the treatment of curved lines, a curved surface can also be divided into N identical parts, each of which is scaled down by a ratio $\delta = 1/\sqrt{N}$ from the whole. The fractal dimension (D) of a curved surface can be calculated similarly to curved lines. The relationship between the resultant area, the size of the ruler used as a measurement unit, and the fractal dimension can be expressed as follows:

$$A = C \times \delta^{2-D} \qquad (2.10)$$

where A is the area of the curved surface, C is a constant, and δ is the size of the area used as a basic unit of measurement. Suppose the sizes of the area units are $\delta_1, \delta_2, \delta_3, \ldots,$ and δ_M; the corresponding measured areas are $A_1, A_2, A_3, \ldots, A_M$. The formula to compute the fractal dimension of the curved surface can be expressed as follows (Wang and Wu, 1998):

$$D = 2 - \frac{\sum_{i=1}^{M} [\log(\delta_i)\log(A_i)] - \frac{1}{M} \times \left(\sum_{i=1}^{M} \log(\delta_i)\right) \times \left(\sum_{i=1}^{M} \log(A_i)\right)}{\left(\sum_{i=1}^{M} \log(A_i)\right)^2 - \frac{1}{M} \times \left(\sum_{i=1}^{M} \log(A_i)\right)^2} \qquad (2.11)$$

where M is the number of units used for measurements.

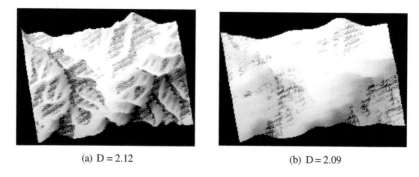

(a) D = 2.12   (b) D = 2.09

FIGURE 2.11 Two surfaces with different dimensions.

Figure 2.11 shows two surfaces with different dimensions: 2.12 and 2.09. The right one is the generalized result of the left one. It is clear that the complexity is much reduced so that the fractal dimension is also smaller.

2.2 MATHEMATICAL MORPHOLOGY*

Mathematical morphology is a science of form and structure based on set theory. It was developed by French geostatistical scientists G. Matheron and J. Serra in the 1960s (Matheron, 1975; Serra, 1982). Since then it has found increasing applications in digital image processing. Mathematical morphology has been employed for the algorithms and operators of transformations in scale dimension (Li, 1994; Su and Li, 1995; Li and Su, 1996; Su et al., 1996, 1997a, 1997b, 1998).

2.2.1 BASIC MORPHOLOGICAL OPERATORS

Because *morphological operators* are based on *set operators*, it is necessary to have a brief outline of set operators in this subsection.

A spatial feature can be considered as a set. Then set operators can be employed to manipulate spatial features. These operators include *union* ($\cup$), *intersection* ($\cap$), *difference* (−), and so on. Figure 2.12 illustrates the manipulation of two area features (A and B) by these set operators.

There are two basic operators developed in *mathematical morphology*: *dilation* and *erosion*. These two operators can be compared to +, −, ×, and ÷ in

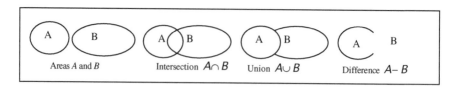

FIGURE 2.12 Manipulation of two spatial features by set operators.

---

* This section is largely extracted from Su et al., 1998. With permission.

ordinary algebra and they are defined as follows (see Serra, 1982; Haralick et al., 1987):

$$\text{Dilation:} \quad A \oplus B = \{a+b: \ a \in A, b \in B\} = \bigcup_{b \in B} A_b \qquad (2.12)$$

$$\text{Erosion:} \quad A \ominus B = \{a: \ a+b \in A, b \in B\} = \bigcap_{b \in B} A_b \qquad (2.13)$$

where A is a feature set (or image) to be processed, and B is another set, called the *structuring element*. Equation (2.12) represents dilation of A by B and Equation (2.13) represents erosion of A by B.

These two operators are illustrated in Figure 2.13, where the features are represented by gray pixels marked with 1. Figure 2.13a is the original feature (image) A. Figure 2.13b is the structuring element B. The origin of a structuring element is considered to be its geometric center if there is no other specific indication. Figure 2.13c is the result of feature A dilated by B. The symbol + indicates pixels that are a result of this dilation process. In ordinary terms, this dilation is achieved by moving the structuring element (as a template) along row and column. If a black (or 1) pixel is encountered, the template center is placed on the pixel position and the template is stamped onto this position to produce a set of output pixels. Figure 2.13d is the result of feature A eroded by B. The symbol − indicates pixels that are a result of

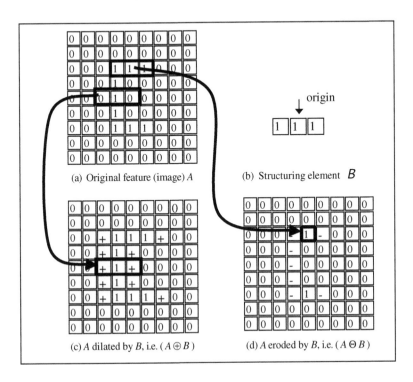

(a) Original feature (image) A

(b) Structuring element B

(c) A dilated by B, i.e. $(A \oplus B)$

(d) A eroded by B, i.e. $(A \ominus B)$

FIGURE 2.13 Two basic morphological operators: dilation and erosion.

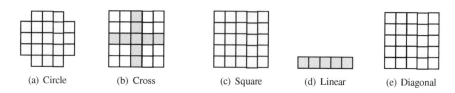

| (a) Circle | (b) Cross | (c) Square | (d) Linear | (e) Diagonal |

FIGURE 2.14 Some possible structuring elements.

this erosion process. This is achieved by moving the template to each pixel to find an exact match. If there is a match, the pixel underneath the center of the template is retained; otherwise, it is eroded, that is, changed to white (or zero).

If a *symmetric structuring element* (e.g., 3×3) is used for dilation and erosion, then the original feature will be dilated or eroded uniformly in all directions. The dilation and erosion in such a special case are called *expansion* and *shrink*, respectively.

The structuring element is a critical element in any morphological operation. A structuring element has three basic parameters: shape, size, and orientation. The result of a morphological operation is completely dependent on these three parameters of a structuring element. Figure 2.14 shows some of the possible shapes: circular, diagonal, linear, square, triangle, and so on. Here, circular structuring element means that it is used to approximate a circle in mathematical morphology. The results of a feature (image) dilated by all of these five types of structuring elements are illustrated in Figure 2.15.

The origins of the structuring elements used in the Figure 2.14 are assumed to be their geometric centers. However, the center of a structuring element could be

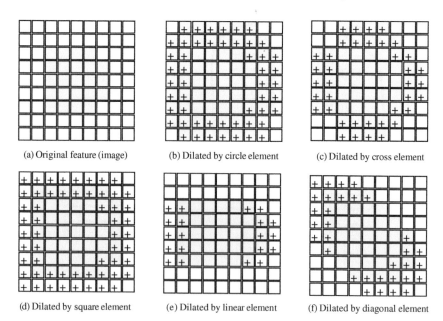

| (a) Original feature (image) | (b) Dilated by circle element | (c) Dilated by cross element |
| (d) Dilated by square element | (e) Dilated by linear element | (f) Dilated by diagonal element |

FIGURE 2.15 Dilation of features by different types of structuring elements.

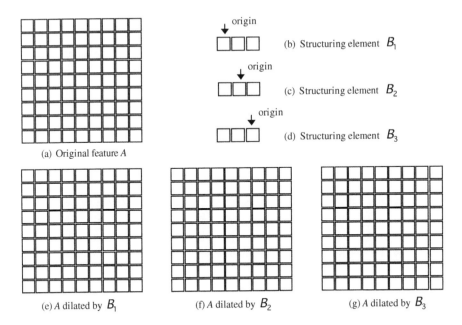

(a) Original feature A

(b) Structuring element B_1

(c) Structuring element B_2

(d) Structuring element B_3

(e) A dilated by B_1

(f) A dilated by B_2

(g) A dilated by B_3

FIGURE 2.16 Dilation of feature A by the same type of structuring element but with different origins.

anywhere else. Figure 2.16 shows the dilation of a feature (image) by the same type of structuring element but with different origins.

A structuring element could be any size, for example, 1×2, 1×3, ..., 2×2, 2×3, ..., 3×3, and so on. The commonly used sizes are 3×3, 5×5, and 7×7. Figure 2.17 shows the dilation of feature (image) A by the same type of structuring element but with different sizes.

2.2.2 Advanced Morphological Operators

Upon the two basic operations, dilation and erosion, many advanced operators have also been developed, such as opening, closing, thinning, thickening, hit_miss, conditional dilation, conditional erosion, conditional thinning, conditional thickening, sequential dilation, sequential thinning, conditional sequential dilation, and so on. In this text, not all of them will be presented, and readers who are interested in this topic are referred to the book by Serra (1982).

The *opening* and *closing* operators are expressed by Equation 2.14 and Equation 2.15.

$$\text{Opening:} \quad A \circ B = (A \ominus B \oplus B) \tag{2.14}$$

$$\text{Closing:} \quad A \bullet B = (A \oplus B \ominus B) \tag{2.15}$$

where A is the original feature (or image) and B is the structuring element. Examples of these two operators are given in Figure 2.18. Figure 2.18a is the original feature

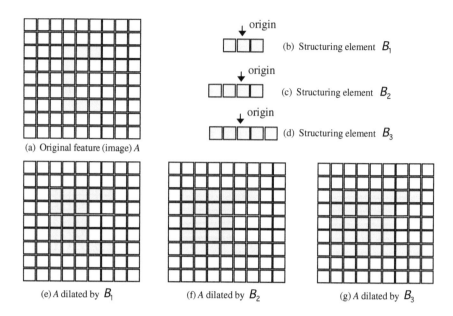

(a) Original feature (image) *A*

↓ origin　(b) Structuring element *B*₁

↓ origin　(c) Structuring element *B*₂

↓ origin　(d) Structuring element *B*₃

(e) *A* dilated by *B*₁　　(f) *A* dilated by *B*₂　　(g) *A* dilated by *B*₃

FIGURE 2.17 Dilation of features by the same type of structuring element but with different sizes.

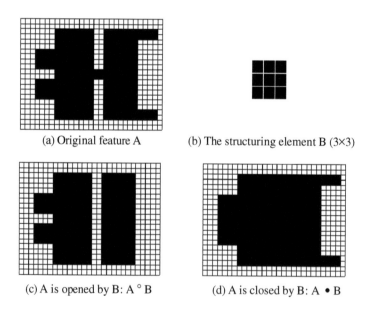

(a) Original feature A　　　(b) The structuring element B (3×3)

(c) A is opened by B: A ° B　　(d) A is closed by B: A • B

FIGURE 2.18 The process of opening and closing operators (Reprinted from Su et al., 1998. With permission.).

consisting of two connected blocks. Figure 2.18c shows the result of opening by
structuring element B, that is, two separated blocks. In other words, the connection
is cut and thus the original feature is opened, because the connection part is thinner
(only 2×2) than the structuring element B (3×3). By contrast, Figure 2.18d shows
that the gaps between the two blocks are filled up, or closed, because the gaps are
all narrower than the structuring elements.

The *hit_miss* operator is the basic tool for shape detection and construction in
morphological algorithms. It is defined as follows:

$$A \otimes B = (A \ominus B_1) \cap (A^C \ominus B_2) \qquad (2.16)$$

where A is an original feature, A^c is a complement (or background) of A, and B is
a structuring element pair, that is, $B = (B_1, B_2)$, one to probe the inside and one to
probe the outside of the feature. A point is obtained in the hit_miss output if and
only if B_1 translated to the point fits inside A and B_2 translated to the point fits
outside A. It is assumed that B_1 and B_2 are disjoint, that is, $B_1 \cap B_2 = \phi$. The process
of the hit_miss operator is illustrated in Figure 2.19. In this case the hit_miss operator
is used to detect and extract a feature with a "Γ" shape. Figure 2.19a shows the
original feature. and Figure 2.19f shows the result. The hit_miss operator is so
important that it is involved in most morphological algorithms.

Thinning is another advanced operator widely used in raster data processing.
The thinning of feature A by a structuring element B can be described by Equation 2.17
(Serra, 1882).

$$A \bigcirc B = A - (A \otimes B) = A \cap (A \otimes B)^C \qquad (2.17)$$

The symbols $\bigcirc$, $-$, $\otimes$, and $\cap$, respectively stand for thinning, difference,
hit_miss, and set intersection. A^c is the complement of set A.

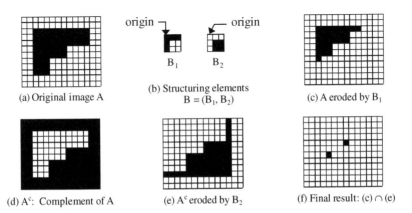

(a) Original image A

(b) Structuring elements
B = (B₁, B₂)

(c) A eroded by B₁

(d) Aᶜ: Complement of A

(e) Aᶜ eroded by B₂

(f) Final result: (c) ∩ (e)

FIGURE 2.19 The process of the hit_miss operator (Reprinted from Su et al., 1998. With
permission.).

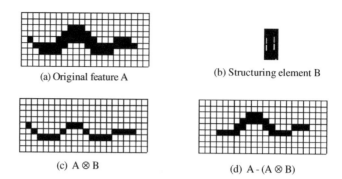

(a) Original feature A   (b) Structuring element B

(c) A ⊗ B   (d) A - (A ⊗ B)

FIGURE 2.20 The process of the thinning operation (Reprinted from Su et al., 1998. With permission.).

The thinning process is illustrated in Figure 2.20. It is clear that, as the name implies, the thinning process makes the area thinner. That is, the result shown in Figure 2.20d is thinner than the original feature shown in Figure 2.20a. The feature shown in Figure 2.20d is not yet the skeleton of the original feature. To obtain a skeleton, the thinning process needs to be repeated so that the final result is a line feature with a width of only one pixel. In practice, a set of structuring elements is used sequentially in this repeated process. In other words, a more useful expression for symmetrical (more than once) thinning of a feature, A, is based on a sequence of structuring elements that are described by Equation 2.18 and illustrated in Figure 2.21.

$$\{B_i\} = \{B_1, B_2, B_3, ..., B_{n-1}, B_n,\} \tag{2.18}$$

Using this sequence of structuring elements, the process of a *systematic thinning* can be described by Equation 2.19 (Serra, 1982):

$$S_k(A) = A \bigcirc (B_i) = ((...((A \bigcirc B_1) \bigcirc B_2)...) \bigcirc B_n \tag{2.19}$$

The sequence of structuring elements shown in Figure 2.21 is called *homotopic structuring elements*. Accordingly, the algorithm described by Equation 2.19 is called the *homotopic sequential thinning* algorithm. (Homotopic sequential thinning is a type of homotopic transformation that has the characteristics of the retention of the original topology.)

The homotopic sequential thinning process described in Equation 2.19 is to thin A by one pass with B_1, then the result with one pass of B_2, and so on, until A is finally thinned with one pass of B_n.

FIGURE 2.21 Sequence of structuring elements $\{B_i\}$ for systematic thinning (x means "don't care").

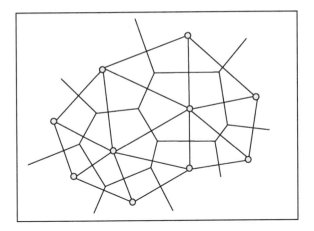

FIGURE 2.22 The dual relationship between Delaunay triangulation and the Voronoi diagram.

2.3 DELAUNAY TRIANGULATION AND THE VORONOI DIAGRAM

Delaunay triangulation and the *Voronoi diagram* have been widely used as data structures to support multi-scale representation (Jones et al., 1995; Ai and Liu, 2002; Li et al., 2004).

These two have a dual relationship, as shown in Figure 2.22. That is, one can be derived from the other with ease. The Voronoi diagram is formed by the bisectors of the triangle sides (edges), and Delaunay triangulation can be formed by joining each pair of points with a common Voronoi edge.

2.3.1 DELAUNAY TRIANGULATION

A huge body of literature on triangulation is available. Detailed discussion of such algorithms lies outside the scope of this text, and good reviews of such algorithms have been written by Tsai (1993), Petrie and Kennie (1990), Weibel and Heller (2000), and Li et al. (2005). This section briefly discusses the general principle.

From a set of randomly distributed data, there are alternative ways to form *triangular irregular networks* (*TINs*). Figure 2.23 illustrates the three alternative

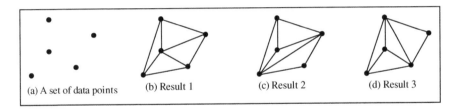

(a) A set of data points   (b) Result 1   (c) Result 2   (d) Result 3

FIGURE 2.23 Triangular networks with different shapes constructed from the same data set.

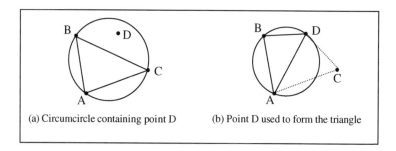

(a) Circumcircle containing point D   (b) Point D used to form the triangle

FIGURE 2.24 The empty circumcircle principle of Delaunay triangulation.

TINs generated from the same set of five data points. The question is, "Which is the most reasonable?" This question can be answered only after some standards or principles have been established.

One of the basic characteristics of Delaunay triangulation is that no other data points are contained by the circumcircle of a *Delaunay triangle*. This is one of the basic principles of the generation of a Delaunay triangulation from a set of 2-D data points and is referred to as the *empty circumcircle principle*, also called the *Delaunay principle* in some literature. Figure 2.24 illustrates this principle. In Figure 2.24a point D is within the circle circumscribing ΔABC if point C is selected as the third vertex to form a triangle with points A and B as the other two vertices. This means that point D instead of point C should be used to form a triangle with points A and B. Figure 2.24b shows this case, where point C does not fall into the circle circumscribing ΔABD.

It is understandable that computing a circumcircle for three given points and checking whether a point is within a circle are not efficient processes. Therefore, some optimization procedures are desirable if they are of help in fulfilling the Delaunay principle. The *local optimization procedure* (LOP) by Lawson (1972) serves such a purpose. It says that the triangular network is optimum if for every convex quadrilateral formed by two adjacent triangles, the swapping of diagonals will not cause a decrease in the minimum of the six interior angles concerned and at the same time will not cause an increase in the maximum angle. In this way, the minimum angle is maximized and the maximum angle is minimized for all the triangles. This is also called the MAX-MIN angle principle. Figure 2.25 illustrates this principle. In Figure 2.25a, two triangles, ΔABC and ΔADC, are used to form a convex quadrilateral. The minimum interior angle is ∠CAD and the maximum interior angle is ∠ADC. After swapping the diagonal, as shown in Figure 2.25, the minimum interior angle then becomes ∠CBD, which is larger than ∠CAD, and the maximum interior angle is ∠ADB, which is smaller than ∠ADC. This means that the shape in Figure 2.25 is the optimal configuration.

2.3.2 CONSTRAINED DELAUNAY TRIANGULATION

Delaunay triangulation with consideration of a given constraint is *constrained Delaunay triangulation* (CDT). A CDT is not truly a Delaunay triangulation, as

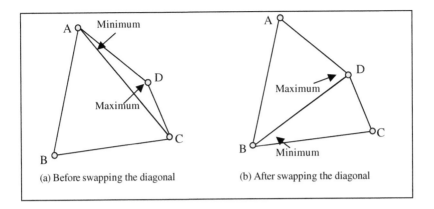

(a) Before swapping the diagonal   (b) After swapping the diagonal

FIGURE 2.25 Illustration of the LOP process for local equiangularity.

some of its triangles might not follow the empty circumcircle principle. For a given set of data points and a set of lines as constraints, CDT is the triangulation of vertices with the following properties:

- The given constraint lines are included in the triangulation.
- The resulting triangulation is as close as possible to the Delaunay triangulation.

By *constrained*, it is meant that the predefined lines are not to be crossed by any triangle edges. To accommodate this, the empty circle principle is modified to apply only to points that can be seen from at least one edge of the triangle where the predefined lines are treated as opaque. As a result, the principle of CDT becomes that *only when the circumcircle of the triangle does not contain any other points, and its three vertexes are visible to each other, this triangle is a constrained Delaunay triangle.*

Here, visibility plays a central role. Figure 2.26 illustrates the intervisibility of data points after the insertion of a predefined line (i.e., constraint).

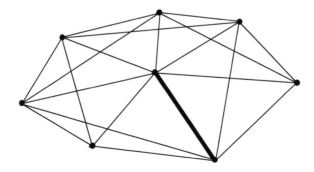

FIGURE 2.26 Intervisibility of eight points and one constrained line segment.

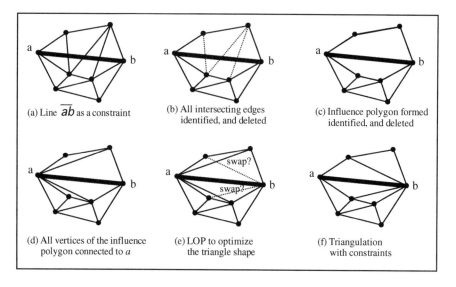

(a) Line $\overline{ab}$ as a constraint

(b) All intersecting edges identified, and deleted

(c) Influence polygon formed identified, and deleted

(d) All vertices of the influence polygon connected to a

(e) LOP to optimize the triangle shape

(f) Triangulation with constraints

FIGURE 2.27 Constrained Delaunay triangulation.

A two-step procedure is commonly used for the construction of CDT as follows:

- Construct the standard Delaunay triangulation with all the data points, including those on the predefined line segments (called constraint line segments).
- Embed the constraint line segments and adjust all triangles in the local areas where they exist through the diagonal swapping process using LOP.

Figure 2.27 shows the process. In Figure 2.27a the standard Delaunay triangulation is completed and constraint line ab is inserted. In Figure 2.27 the triangles whose edges intersect the constraint line segment are identified (Figure 2.27b). If two such triangles have a common edge, this edge is deleted. In this way, the so-called influence polygon of the constraint line segment is formed (Figure 2.27c). All vertices of the influence polygon are connected to the starting point of the constraint line segment, that is, point a (Figure 2.27d). The Lawson LOP is then applied to optimize the local triangles, but the constraint line segment is still an edge of some triangles (Figure 2.27e). The final result is shown in Figure 2.27f.

2.3.3 Voronoi Diagram

A huge body of literature on computation of Voronoi diagrams is available (Green and Sibson, 1977; Brassel and Reif, 1979; Bowyer, 1981; Lee and Drysdale, 1981; Ohya et al., 1984; Klein, 1988; Okabe et al., 1992). Efforts have also been made on the development of dynamic Voronoi diagrams (Zaninetti, 1990; Gold and Condal, 1995), on Voronoi diagrams in raster mode (Li et al., 1999), and on spherical features (Chen et al., 2003). Again, only the general principle is described in this section; more detailed discussions can be found in the books by Okabe et al. (1992) and Li et al. (2005).

From the viewpoint of computational geometry, a Voronoi diagram is essentially a partition of the plane into N polygonal regions, each of which is associated with a given point. The region associated with a point, called the *Voronoi region* or *Thiessen polygon* (Thiessen, 1911), is "the set of points closer to that point than to any other given point" (Lee and Drysdale, 1981). Suppose there are N distinct points P_1, P_2, ..., P_n in the plane. Each point will have a Thiessen polygon. All these Thiessen polygons together form a pattern of packed convex polygons covering the whole plane (no gap or overlap). That is,

$$V = \{V(p_1), V(p_2), \cdots, V(p_n)\} \qquad (2.20)$$

where V is Voronoi diagram of the data set and $V(p_i)$ denotes the Voronoi region of a point feature p_i.

A simple method for constructing a Voronoi diagram is the incremental method (Green and Sibson, 1977; Bowyer, 1981; Lee and Drysdale, 1981; Ohya et al., 1984). The basic idea of the incremental method is to expand the Voronoi diagram incrementally, that is, to add one point at a time. Figure 2.28 illustrates the principle. First, the Voronoi diagram of the first three points, 1, 2 and 3, is constructed, and point 4 is to be considered. It is found that point 4 is located within the Voronoi region of point 1. Then a perpendicular bisector between points 1 and 4 is drawn. This perpendicular bisector intersects the common boundary of Voronoi regions $V(p_1)$ and $V(p_3)$. Then a perpendicular bisector between points 4 and 3 is drawn. As a result, part of $V(p_1)$ and part of $V(p_3)$ together form $V(p_4)$. The process continues until the Voronoi diagram of N points is computed through the Voronoi diagram of $N-1$ points, by adding the last point.

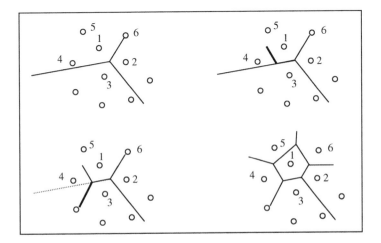

FIGURE 2.28 Incremental method for computing a Voronoi diagram.

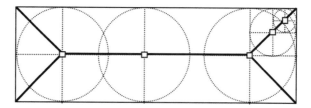

FIGURE 2.29 Skeleton formed by a locus of the center of a disc touching the boundary.

2.4 SKELETONIZATION AND MEDIAL AXIS TRANSFORMATION

The terms *medial axis transform* (*MAT*) and *skeletonization* are often used interchangeably. The notion of a *skeleton* was introduced by H. Blum (1967) as a result of the MAT. Skeletonization is a process used to produce a skeleton for a given feature. Skeletonization has been employed for the algorithms and operators of geometric transformations in multi-scale representation (e.g., Su et al., 1998; Christensen, 1999, 2000). Many skeletonization algorithms are available (e.g., Bookstein, 1979). Skeletonization can be produced via MAT, Voronoi diagram, and thinning.

2.4.1 SKELETONIZATION BY MEANS OF MAT AND DISTANCE TRANSFORM

The MAT, as the name implies, is a process used to derive the lines formed by midpoints. These midpoints form the skeleton of the feature and are the closest boundary point(s) for each point in a feature. A skeleton point has at least two closest boundary points. Each point on the MAT of a continuous shape is the center of a disc touching the boundary at two (or more) locations. Another way to think about the skeleton is as the loci of centers of bitangent circles that fit entirely within the foreground region (of a feature) being considered. Figure 2.29 shows the skeleton of a rectangle by MAT.

In raster mode, the MAT is achieved by a *distance transform* of the raster image, which means computing the raster distance of each image pixel to the feature pixels (Tang, 1989). For each pixel, the smallest among all possible distances is taken. Figure 2.30 shows the distance transform of a rectangle, which is represented by

0	0	0	0	0	0	0	0	0	0	0	0	0	0	0	0	0	0	0	0	0	0	0	0	0	0	0	0	0	0
0	1	1	1	1	1	1	1	1	1	1	1	1	1	1	1	1	1	1	1	1	1	1	1	1	1	1	1	1	0
0	1	2	2	2	2	2	2	2	2	2	2	2	2	2	2	2	2	2	2	2	2	2	2	2	2	2	1	0
0	1	2	3	3	3	3	3	3	3	3	3	3	3	3	3	3	3	3	3	3	3	3	3	3	2	1	0
0	1	2	3	4	4	4	4	4	4	4	4	4	4	4	4	4	4	4	4	4	4	4	3	2	1	0
0	1	2	3	3	3	3	3	3	3	3	3	3	3	3	3	3	3	3	3	3	3	3	3	2	1	0
0	1	2	2	2	2	2	2	2	2	2	2	2	2	2	2	2	2	2	2	2	2	2	2	2	1	0
0	1	1	1	1	1	1	1	1	1	1	1	1	1	1	1	1	1	1	1	1	1	1	1	1	1	0
0	0	0	0	0	0	0	0	0	0	0	0	0	0	0	0	0	0	0	0	0	0	0	0	0	0	0

FIGURE 2.30 Distance transform for skeletonization.

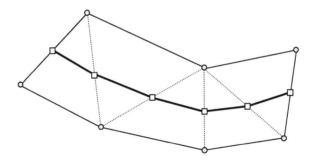

FIGURE 2.31 Derivation of a skeleton by triangulation.

boundary pixels with 0. That is, the distance values of all these feature pixels to the boundary of the area feature are zero. The distance transform in this figure is produced using the chessboard distance. In the end, the skeleton is very clear and consists of those pixels with raster distance 4 (pixels). The result is similar to that shown in Figure 2.29.

2.4.2 Skeletonization by Means of Voronoi Diagram and Triangulation

Another way of generating a skeleton is to construct a triangular irregular network (TIN) of the points on the feature and then to join the middle points of the triangle sides (Jones et al., 1995). Figure 2.31 illustrates the derivation of a skeleton from a TIN.

Another way to derive the skeleton is to make use of the Voronoi diagram, that is, to connect the edges of neighboring Voronoi regions (Gold and Snoeyink, 2001). Figure 2.32a shows such a relationship. The highlighted Voronoi edges in this figure are regarded as the skeleton of this area. However, this skeleton appears to be not smooth enough. To achieve better smoothness, more points should be added to the boundary of the area. Figure 2.32b shows the improvement. Rigorously speaking, one should say that if the density of boundary points goes to infinity, then the corresponding Voronoi diagram converges to the skeleton.

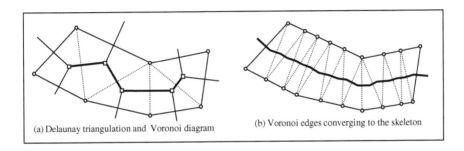

(a) Delaunay triangulation and Voronoi diagram   (b) Voronoi edges converging to the skeleton

FIGURE 2.32 Skeleton formed by Voronoi edges generated from boundary dense points.

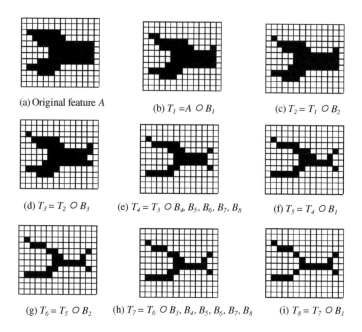

(a) Original feature A

(b) $T_1 = A \ominus B_1$

(c) $T_2 = T_1 \ominus B_2$

(d) $T_3 = T_2 \ominus B_3$

(e) $T_4 = T_3 \ominus B_4, B_5, B_6, B_7, B_8$

(f) $T_5 = T_4 \ominus B_1$

(g) $T_6 = T_5 \ominus B_2$

(h) $T_7 = T_6 \ominus B_3, B_4, B_5, B_6, B_7, B_8$

(i) $T_8 = T_7 \ominus B_1$

FIGURE 2.33 Homotopic sequential thinning algorithm for skeletonization (Reprinted from Su et al., 1998. With permission.).

2.4.3 Skeletonization by Means of Thinning

Skeletonization can also be considered as an erosion process that preserves an important characteristic of feature shapes. The skeleton is produced by an iterative erosion process that removes all pixels that are not part of the skeleton.

In raster, *morphological thinning* is the typical example that successively erodes away pixels from the boundary (while preserving the end points of line segments) until no more thinning is possible. Su et al (1998) have used morphological thinning for multi-scale representation.

The homotopic sequential thinning process was described in Equation 2.19. It is used to thin feature A by one pass with B_1, then thin the result with one pass of B_2, and so on, until A is finally thinned with one pass of B_n. For skeletonization, the entire process is repeated until no further changes occur, that is, when lines with only one pixel width are obtained. This process is illustrated in Figure 2.33.

(a)	(b)	(c)	(d)	(e)	(f)	(g)	(h)

FIGURE 2.34 Sequential structuring elements $\{P_i\}$ for pruning parasitic branches (x means "don't care").

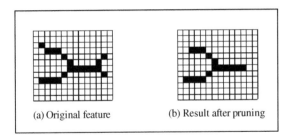

| (a) Original feature | (b) Result after pruning |

FIGURE 2.35 Effect of pruning (Reprinted from Su et al., 1998. With permission.).

As shown in Figure 2.33i, some parasitic branches still exist in the result obtained from this thinning-based skeletonization. They need to be removed. This removal process is called *pruning*, which can be described by Equation 2.21.

$$P(X) = X \bigcirc \{P_i\} \qquad (2.21)$$

The form of this equation is essentially identical to the homotopic sequential thinning algorithm defined by Equation 2.19. The only difference is that a set of specially designed structuring elements as shown in Figure 2.34, $\{P_i\}$, is used, so that this process is capable of pruning parasitic branches in the eight basic directions: right (a), left (c), up (d), down (b), upper right (h), upper left (g), lower right (e), and lower left (f). The effect of pruning is illustrated in Figure 2.35.

REFERENCES

Ai, T. and Liu, Y., A method of point cluster simplification with spatial distribution properties preserved, *Acta Geodaetica et Cartograpgica Sinica*, 25(1), 35–41, 2002. (in Chinese)

Blum, H., A transformation for extracting new descriptors of shape, in *Models for the Perception of Speech and Visual Form*, Whaten Dunn, W., Ed., MIT Press, Cambridge, MA, 1967, pp. 153–171.

Bookstein, F. L., The line-skeleton, *Computer Graphics and Image Processing*, 11, 123–137, 1979.

Borgefors, G., Distance transformations in digital images, *Computer Vision, Graphics and Image Processing*, 34, 344–371, 1986.

Borgefors, G., Applications using distance transforms, in *Aspects of Visual Form Processing*, Arcelli, C., Cordelia, L., and Sanniti di Baja, G., Eds., World Scientific, Singapore, 1994, pp. 83–108.

Bowyer, A., Computing Dirichlet tessellations, *Computer Journal*, 24(2), 162–166, 1981.

Brassel, K. and Reif, D., Procedure to generate Thiessen polygon, *Geographical Analysis*, 11(3), 289–303, 1979.

Breu, H., Gil, J., Kirkpatrick, D., and Werman, M., Linear time Euclidean distance transformation methods, *IEEE Transactions Pattern Analysis and Machine Intelligence*, 17(5), 529–533, 1995.

Buttenfield, B., Scale-dependency and self-similarity in cartographic lines, *Cartographica,* 26(1), 79–99, 1989.

Carstensen, L. W. Jr., A fractal analysis of cartographic generalization, *American Cartographer,* 16(3), 181–189, 1989.

Chen, J., X.S. Zhao, and Li, Z.L., An algorithm for the generation of Voronoi diagrams on the sphere based QTM, *Photogrammetric Engineering and Remote Sensing,* 69(1), 79–90.

Christensen, A. H., Cartographic line generalization with waterlines and medial-axes, *Cartography and Geographic Information Science,* 26(1), 19–32, 1999.

Christensen, A. H., Line generalization by waterline and medial-axis transformation: success and issues in an implementation of Perkel's proposal, *Cartographic Journal,* 26(1), 19–32, 2000.

Embrechts, H. and Roose, D., A parallel Euclidean distance transformation method, *Computer Vision and Image Understanding,* 63(1), 15–26, 1996.

Gold, C. M. and Condal, A. R., A spatial data structure integrating GIS and simulation in a marine environment, *Marine Geodesy,* 18, 213–228, 1995.

Goodchild, M. F., Fractal and accuracy of geographical measures, *Mathematical Geology,* 12(2), 85–98, 1980.

Goodchild, M. and Mark, D., Fractal nature of geographic phenomena, *Annals of the Association of American Geographers,* 7(2), 265–278, 1987.

Green, P. J. and Sibson, R., Computing Dirichlet tessellations in the plane, *Computer Journal,* 21(2), 168–173, 1977.

Haralick, R., Sternberg, S., and Zhuang, X., Image analysis using mathematical morphology, *IEEE Transactions of Pattern Analysis and Machine Intelligence,* 9(4), 532–550, 1987.

Jones, C. B., Bundy, G. L., and Ware, J. M., Map generalization with a triangulated data structure, *Cartography and Geographic Information Systems,* 22(4), 317–331, 1995.

Klein, R., *Abstract Voronoi Diagrams and Their Applications,* Lecture Notes in Computer Science 333, Springer-Verlag, Berlin, 1988.

Kong, T. Y. and Rosenfeld, A., Digital topology: an introduction and a survey, *Computer Vision, Graphics and Image Processing,* 48, 357–393, 1989.

Lawson, C. L., Transforming triangulations, *Discrete Mathematics,* 3, 365–372, 1972.

Lee, D. T. and Drysdale, R. L., Generalization of Voronoi diagram in the plane, *SIAM Journal of Computing,* 10, 73–87, 1981.

Lee, Y. C. and Li, Z. L., A taxonomy of 2D space tessellation, *International Archives for Photogrammetry and Remote Sensing,* 32(4), 344–346, 1998.

Li, Z. L., Mathematical morphology in digital generalization of raster map data, *Cartography,* 23(1), 1–10, 1994.

Li, Z. L. and Su, B., Algebraic models for feature displacement in the generalisation of digital map data using morphological techniques, *Cartographica,* 32(3), 39–56, 1996.

Li, Z. L., Yan, H., Ai, T., and Chen, J., Automated building generalization based on urban morphology and gestalt theory, *Journal of Geographical Information Science,* 18(5), 513–534, 2004.

Li, Z. L., Zhu, Q., and Gold, C., *Digital Terrain Modelling: Principles and Methodology.* CRC Press, Boca Raton, FL, 2005.

Maffini, G., Raster versus vector data encoding and handling: a commentary, *Photogrammetric Engineering and Remote Sensing,* 53(10), 1397–1398, 1987.

Maguire, D. J., Kimber, B., and Chick, J., Integrated GIS: the importance of raster, Technical Papers, *ACSM–ASPRS Annual Convention,* 4, 107–116, 1991.

Mandelbrot, B. B., How long is the coast of Britain? *Science*, 156, 636–638, 1967.

Mandelbrot, B. B., *Fractal: Form, Chance and Dimension*. W.H. Freeman, New York, 1977.

Mandelbrot, B. B., *The Fractal Geometry of Nature*. W.H. Freeman, New York, 1983.

Matheron, G., *Random Sets and Integral Geometry*. Wiley, New York, 1975.

Melter, R. A., Some characterisations of city block distance, *Pattern Recognition Letters*, 6, 235–240, 1987.

Monmonior, M., Raster-mode area generalisation for land use and land cover maps, *Cartographica*, 20(4), 65–91, 1983.

Muller, J.-C., Fractal dimension and inconsistencies in cartographic line representation, *Cartographic Journal*, 23, 123–130, 1986.

Ohya, T., Iri, M., and Murota, K., Improvement of the incremental method for the Voronoi diagram with computational comparison of various methods, *Journal of the Operations Research Society of Japan*, 27, 306–336, 1984.

Okabe, A., Boots, B., and Sugihara, K., *Spatial Tessellations: Concepts and Applications of Voronoi Diagrams*, John Wiley, Chichester, UK, 1992.

Petrie, G. and Kennie, T., Eds., *Terrain Modelling in Surveying and Civil Engineering*, Caitness, UK, Whittles, 1990.

Peuquet, D. J., A conceptual framework and comparison of spatial data models, *Cartographica*, 21, 66–113, 1984. Reprinted in *Introduction Readings in Geographic Information Systems*, Peuquet, D. J. and Marble, D. F., Eds., Taylor & Francis, London, 251–285, 1990.

Richardson, L. F., The problem of contiguity: an appendix to statistics of deadly quarrels, *General Systems Yearbook*, 6, 139–187, 1961.

Rosenfeld, A. and Pfaltz, J., Distance functions on digital pictures, *Pattern Recognition*, 1(1), 33–61, 1968.

Serra, J., *Image Processing and Mathematical Morphology*. Academic Press, New York, 1982.

Su, B. and Li, Z. L., An algebraic basis for digital generalisation of area-patches based on morphological techniques, *Cartographic Journal*, 32(2), 148–153, 1995.

Su, B., Li, Z. L. and Lodwick, G., Morphological transformation for the elimination of area features in digital map generalization, *Cartography*, 26(2), 23–30, 1997a.

Su, B., Li, Z. L., and Lodwick, G., Algebraic models for collapse operation in digital map generalization using morphological operators, *GeoInformatica*, 2(4), 359–382, 1998.

Su, B., Li, Z. L., Lodwick, G., and Müller, J. C., Algebraic models for the aggregation of area features based upon morphological operators, *International Journal of Geographical Information Science*, 11(3), 233–246, 1997b.

Tang, L., Surface modelling and visualization based upon digital image processing techniques, in *Optical 3-D Measurement Techniques*, Grun, A. and Kahmen, H. K., Eds., Wichmann Verlag, Karlsruhe, Germany, 1989, pp. 317–325.

Thiessen, A. H., Precipitation averages for large areas, *Monthly Weather Review*, 39, 1082–1084, 1991.

Tsai, V. J. D., Delaunay triangulations in TIN creation: an overview and a linear-time algorithm, *International Journal of Geographical Information Systems*, 7(6), 501–524, 1993.

Wang, Q. and Wu, H., *Fractal Description of Geographic Phenomena and Automated Map Generalization*, WTUSM Press, Wuhan, China, 1998..

Weibel, R. and Heller, M., Digital terrain modelling, in *Geographical Information Systems: Principles and applications*, Maguire, D. J., Goodchild, M. F., and Rhind, D. W., Eds., Longman, London, 1990, pp. 269–297.

Zaninetti, L., Dynamic Voronoi tessellation, *Astronomy and Astrophysics*, 233, 293–300, 1990.

3 Theoretical Background

Algorithms are developed based on certain principles and theories. To facilitate the presentation of algorithms, a discussion of these principles and theories is provided in this chapter.

3.1 SCALE IN GEOGRAPHICAL SPACE

Scale is a term not well defined yet. "Of all words that have some degree of specialised scientific meaning, 'scale' is one of the most ambiguous and overloaded" (Goodchild and Quattrochi, 1997). In different contexts, it may mean different things. To discuss the scale issues in spatial representation, it is necessary to conduct a discussion on the theory of scale.

3.1.1 GEO-SCALE IN THE SCALE SPECTRUM

Scale may mean different things in different contexts. That is, different types of scale may be differentiated with different criteria. Table 3.1 is an attempt to classify scale based on different criteria.

Digital scale is the scale used in the expression of digital number. For example, we may express the same distance in a number of ways, such as 1.68×10^2 for 168.0 and 1.68×10^{-2} for 0.0168.

Temporal scale is related to the time interval. It may range from a few nanoseconds to hours, days, seasons, years, and even billions of years.

Radiometric scale is related to the detail level of the brightness of image pixels. A binary scale representation produces a black and white image. An image with 256 gray levels smoothly represents detailed variations in brightness.

Spectral scale is related to frequency, concerning the interval of frequency spectrum. For example, in remote sensing, the *electromagnetic spectrum* is a fundamentally important concept. That is, *electromagnetic radiation* (EMR) has a range of waves with different length, from gamma rays to radio waves. Visible light is only a small band in this electromagnetic (EM) spectrum. Different spectral bands (ranges) from this EM spectrum have been selected for remote sensing. The width of a band can be regarded as the spectral scale used.

Spatial scale is related to space and is the main concern of this text. In space-related science, different disciplines study different natural phenomena. Nuclear physics studies particles at the submolecular level in units of nanometers. This is an extreme at a microscale. In the other extreme, astrophysics studies the planets at an intergalactic level in units of light-years (the distance traveled by light in the period of a year). Such studies are at a macroscale. In the middle, many scientific disciplines study the

TABLE 3.1
Classification of Scale Based on Different Criteria

Criteria	Type of Scales
Domain of interest	Digital, spatial, temporal, spectral, radiometric
Scope of interest	Micro, …➔, geo-scale, …➔, macro
Stage of processing	Reality, data source, sampling, processing, model, representation
Level of measurement	Nominal, ordinal, interval, ratio

planet earth, such as geology, geography, geomatics (surveying and mapping), geomorphology, geophysics, and so on. These sciences relevant to the studies of the earth are called geosciences. Such studies are at a scale called *geoscale* in this book.

By analogy to the EM spectrum, the scale range in science, from microscale to geoscale to macroscale, is termed the *scale spectrum* and is shown in Figure 3.1. Like the visible light band in the EM spectrum, geoscale is a small band in the scale spectrum.

Recently, the word *geospatial* has become popular. This is directly related to the geoscale concept. The word *geospatial* was perhaps first used by the U.S. Geologic Survey in a program called "geospatial data infrastructure." It is normally used to refer to spatial information at geoscale.

In spatial data handling, at different stages (from reality, to data, and to final representation), different scales may be used. The concept scale may be expressed in different ways, such as in nominal scale (e.g., temporal, spatial), order scale (e.g., global, continental, national, provincial, local, etc.), or ratio and interval scale (e.g., 1:10,000).

3.1.2 Measures (Indicators) of Scale

The term *scale* may mean different things in different disciplines. In cartography, scale is the ratio of distance on a map to that on the ground, and this ratio is applicable to all other engineering drawings. The important question to be answered is, "Is there any other meaning implied in the scale of a map?" The answer to this question is directly related to the answer to the following question: "Do maps

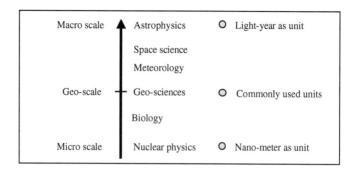

FIGURE 3.1 The scale spectrum and the geoscale.

of the same area at different scales represent the same reality?" The answer to the latter is "no." In other words, maps of the same area but at different scales represent different levels of abstraction. Therefore, scale also implies degree of abstraction or level of detail (LoD), in addition to ratio of distance. If a map is represented in raster, then the size of raster pixels (i.e., *resolution*) is also an indicator of LoD. A map at a given scale also implies a certain level of accuracy, according to map specifications. For a given map area, the size of the ground area varies with scale. The same map area will cover a larger ground area if the map scale is smaller. That is why in geography the size of the study area is used to indicate scale or LoD.

In summary, a set of parameters should be used as the measures (or indicators) of scale:

- Cartographic ratio
- Size of study area (i.e., geographical context)
- Resolution
- Accuracy

It should be emphasized here that scale is meaningful only when these parameters are consistent. Figure 3.2 (See color insert following page 116) shows four images with same image size and ground coverage. Thus, the cartographic ratio of these four images is identical, but they represent LoD because the resolutions are different. Therefore, it can be said that the scale of these four images is different.

Maps in digital form can be plotted at any scale one wishes, but the resolution and accuracy of the digital data is fixed. Therefore, there is no point to plotting a map at a very large scale if the accuracy requirement cannot be met.

The concept of resolution should be more precisely called *spatial resolution*, because there are other resolutions for spatial data (i.e., temporal, spectral, and radiometric resolutions).

3.1.3 Transformations of Spatial Representation in Scale in Geographical Space

In Euclidean space, an increase (or decrease) in scale will cause an increase (or decrease) in length, area, and volume in a three-dimensional (3-D) space. However, the shape and complexity of a feature remain unchanged. Figure 3.3 is an example of scale reduction in a 2-D Euclidean space. The graphic representation of the feature at scale 2 is a 2 times reduction of that at scale 1, and the graphic representation at scale 3 is a 4 times reduction of that at scale 1. In such a transformation process, the (area) size of the graphic representation is reduced by 2^2 and 4^2 times, respectively. When the graphic representation at scale 3 is increased by 4 times, the enlarged graphic is identical to original one shown at scale 1. That is, the transformations are reversible.

However, in the *fractal geographical space*, as discussed in Chapter 2, it was discovered long ago that different length values will be obtained for a coastline represented on maps at different scales. The length measured from a smaller scale

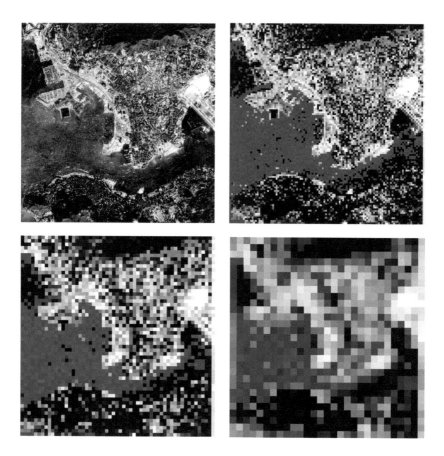

FIGURE 3.2 (See color insert following page 116) Four images with the same cartographic ratio but different resolutions.

map will be shorter if the same unit size (at map scale) is used for measurement. This is because a different level of reality (e.g., the Earth's surface with different degrees of abstraction) has been measured. At a smaller scale, the size of the graphical representation is reduced. At the reduced graphics, the complexity of the graphics is still retained at the same level. As a result, if the graphic is enlarged back to the size at the larger scale, the level of complexity of the representation will appear to be much reduced. In other words, in geographical space, the level of complexity cannot be recovered by an increase in scale. Figure 3.4 illustrates the effect of an increase and decrease in scale on spatial representation in geographical space. It shows that the transformations in scale in such geographical space are not reversible.

In this example, in Euclidean space the reduction of graphic representation in size does not cause a change in its complexity in an absolute sense. This can be understood with the following line of thought: When the size of the graphic is changed, the basic resolution of the observational instrument is also changed by

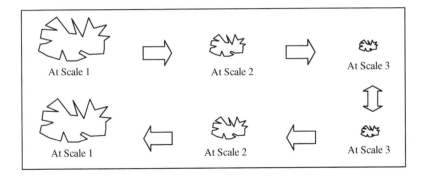

FIGURE 3.3 Scale change in Euclidean space: a reversible process.

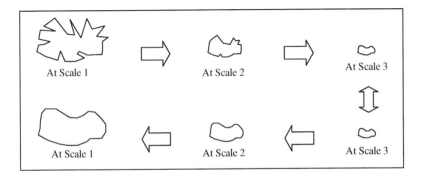

FIGURE 3.4 Scale change in 2-D geographical space: lost complexity is not recovered.

the same magnification. However, the reduction of a graphic in size does cause a change in complexity in a relative sense. It is clear from Figure 3.3 that the graphic at scale 3 appears more complex than other two. This is because these three graphics are observed by the same observer, that is, with an identical resolution (Table 3.2).

TABLE 3.2
Causes and Effects of Scale Reduction in Euclidean and Geographical Space

	Effect		Cause	
Space	Relative Complexity	Absolute Complexity	Instrumentation Resolution	Observer's Resolution
Euclidean space	Increased	Unchanged	Reduced	Unchanged
Geographical space	Unchanged	Decreased	Unchanged	Unchanged

However, in geographical space, the change of complexity is achieved by changing the relationship between the size of the feature and the basic resolution of the observation instrument. There are two ways to achieve this result. The first is to change the size of the feature but to retain the basic resolution of the observational instrument (Table 3.2). The second way is (a) to retain the size of the feature unchanged but to change the basic resolution of the observation instrument, and then (b) to change the size of observed features by a simple reduction in Euclidean space.

3.2 RELATIVITY IN SCALE: THE NATURAL PRINCIPLE

It is a common sense that when taking a picture, you see more detail if you zoom in, and you see less detail if you zoom out. This is a principle underlining this natural phenomenon and it is a basis for multi-scale spatial representation.

3.2.1 THE IDEA OF A NATURAL PRINCIPLE

Reality at different scales means different things, as discussed in Chapter 1. A simple example is the Earth's surface viewed from different heights, which was used by Li and Openshaw (1993). If one views the terrain surface from a satellite, it becomes very smooth. When one views the terrain surface from an airplane, small details do not appear and the main characteristics of the terrain variations are very clear. These are just some of the many practical examples illustrating the transformation in scale dimension. In such a transformation the (absolute) complexity of spatial features has been altered with a change in scale (Li, 1996).

This is also due to the limitation of the human eye's resolution. When the viewpoint is higher, the ground area corresponding to the human eye's resolution becomes larger, and thus the ground surface appears to be more abstract. In the case of stereo models formed from images, it is due to the resolution of images. That is, all information within the image resolution (e.g., 10 m per pixel in the case of SPOT images) disappears.

These examples underline a universal principle, a *natural principle* as it is called by Li and Openshaw (1993), which states:

> For a given scale of interest, all details about the spatial variations of geographical objects (features) beyond a certain limitation cannot be presented and can thus be neglected.

It follows, therefore, that a simple corollary to this process can be used as a basis for the transformations in scale dimension. The corollary can be stated as follows:

> By using a criterion similar to the limitation of the human eye's resolution and, neglecting all the information about the spatial variation of spatial objects (features) beyond this limitation, zooming (or generalization) effects can be achieved.

Li and Openshaw (1992) call such a limitation the *smallest visible object* (SVO), called *smallest visible size* (SVS) in other literature. Figure 3.5 illustrates

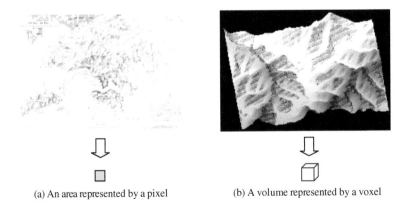

(a) An area represented by a pixel   (b) A volume represented by a voxel

FIGURE 3.5 The natural principle. Spatial variations in an SVS can be neglected.

the natural principle to generate a zooming effect for a 2-D (left) and a 3-D representation (right).

Figure 3.6 illustrates the working example of applying this natural principle to 2-D representations. In the upper-left of this diagram is a river depicted by its two banks. On the upper right is a template full of overlapping SVSs. The SVS represents the size for representation at a smaller scale but has been enlarged to match the map scale of the river. On the lower left is the overlay of this river feature onto the template. The lower right shows the result at the smaller scale where every SVS becomes a point (i.e., represented by a dot in this diagram). This principle mimics

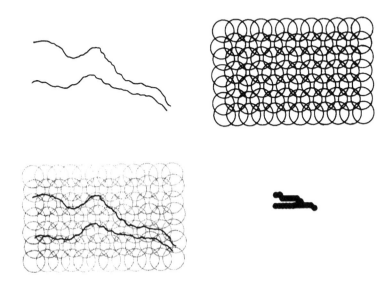

FIGURE 3.6 The natural principle applied to a 2-D representation (Reprinted from Li and Openshaw, 1993).

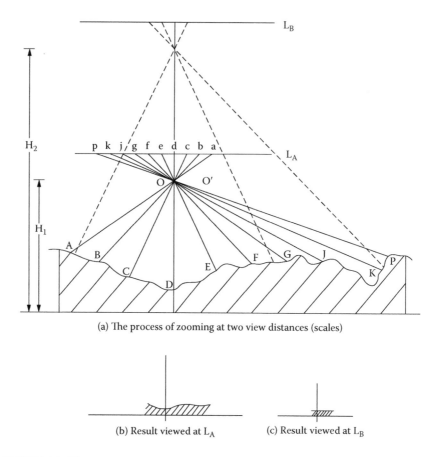

(a) The process of zooming at two view distances (scales)

(b) Result viewed at L_A (c) Result viewed at L_B

FIGURE 3.7 The natural principle applied to a 3-D representation (Reprinted from Li and Openshaw, 1993).

the effect of zooming when a photograph is taken. Li (1996) refers to this *zooming effect* as the transformation in scale dimension.

Figure 3.7 illustrates the working example of applying this natural principle to 3-D representations. Figure 3.7a shows the view of a 3-D surface at two different heights, resulting in representations at two different scales. Figure 3.7b shows the result viewed at level L_A, and Figure 3.7c shows the result viewed at level L_B. In these latter two figures, the zooming effects are very clear.

3.2.2 ESTIMATION OF PARAMETERS FOR THE NATURAL PRINCIPLE

To apply this natural principle, the critical element to be considered is the value of this "certain limitation," that is, the value of the SVS, beyond which all spatial variations (no matter how complicated) can be neglected. This is related to the thresholds of perception in visual science and to map scales. That is, the minimum separation and minimum size of symbols on maps may be used as a reference.

TABLE 3.3
Minimum Size Required for Various Types of Map Symbols

Map Symbols	Type of Symbols	Thresholds
●	Point	0.2 mm
———	Line	Thickness: 0.1 mm
■/□	Square	Side: 0.4 mm/0.6 mm
●/○	Circle	Diameter: 0.4 mm/0.5 mm
▲/△	Equalateral triangle	Width: 0.6 mm/0.7 mm
▲/△	Isosceles triangle	Width: 0.4/0.5 mm, Height: 1.0 mm
▬/▭	Rectangle	Width: 0.4 mm/0.5 mm

It is generally understood that the minimum separation between two map symbols is 2 mm at map scale. The minimum sizes of map symbols are listed in Table 3.3, which is extracted from Spiess (1988). From this table it is can be seen that the thresholds of point symbols range from 0.2 mm to 1.0 mm. Therefore, the value of the SVS for any spatial representations might be between 0.2 mm and 1.0 mm.

Theoretically speaking, the absolute value of the SVS on the ground (K) must be a function of the SVS value (k) on the map (or other spatial representation) and the scale ($1{:}S_T$) of the target map (or other spatial representations). A natural thought is

$$K = k \times S_T \tag{3.1}$$

However, there is a problem associated with this formula. That is, the K value is the same no matter what the scale of the source map. To solve this problem, Li and Openshaw (1992, 1993) modified Equation 3.1 as follows:

$$K = k \times S_T \times \left(1 - \frac{S_S}{S_T}\right) \tag{3.2}$$

where S_T and S_S are the scale factors of the target map and source maps, respectively, and k is the SVS value in terms of map distance on the target map. Through intensive experimental testing, Li and Openshaw (1992) found that a value between 0.5 mm and 0.7 mm enabled them to produce line generalization results similar to manual generalization. Therefore, it is recommend that

$$k = \{0.5mm, 0.7mm\} \tag{3.3}$$

Equation 3.2 seems more reasonable. When the difference between S_S and S_T is small, the K value is small. This means that not much is to be changed in the transformations. In an extreme, when $S_T = S_S$, $K = 0$. This means nothing should be changed in the transformations.

3.3 THE RADICAL LAWS: PRINCIPLES OF SELECTION

In the cartographic community, some interesting studies have been conducted on the relationship between the numbers of symbols at different scales of representation. An empirical law has been formed from these studies, called the *principle of selection*. This principle has been used to determine how many symbols to retain on a representation at a smaller scale.

3.3.1 NUMBER OF SYMBOLS AT DIFFERENT SCALES: A THEORETICAL ANALYSIS

It is clear that for the same ground area, if a map (at a smaller scale) is being derived from maps at a larger scale, the map space is reduced. As a result, not all the symbols on the maps at a larger scale can be represented on the map at a smaller scale. That is, the absolute number of total symbols must be reduced on the small-scale map. However, the relative number of symbols in terms of per unit of map area should be approximately retained. In other words, the density of symbols on map should be retained somehow. Mathematically,

$$\frac{N_S}{A_S} \approx \frac{N_T}{A_T} \tag{3.4}$$

where N_S and N_T are the numbers of symbols on the source and target maps, respectively, and A_S and A_T are the areas of the maps at source and target scales, respectively.

Suppose that a ground area is $L_2 \times L_2$. The corresponding areas on source map A_S and on target map A_T are as follows:

$$A_S = \frac{L_1}{S_S} \times \frac{L_2}{S_S}$$
$$A_T = \frac{L_1}{S_T} \times \frac{L_2}{S_T} \tag{3.5}$$

where S_S and S_T are the same as in Equation 3.2. By substituting Equation 3.5 into Equation 3.4, the following equation can be obtained:

$$N_T \approx N_S \times \frac{S_S^2}{S_T^2} \approx N_S \times \left(\frac{S_S}{S_T}\right)^2 \tag{3.6}$$

This is called the *equal map density* function. This is the case when the minimum size of cartographic symbols is not considered. When considering the need to exaggerate some cartographic symbols on a map at a smaller scale, Equation 3.6 needs to be modified. That is, one or more adjustment factors need to be introduced into Equation 3.6.

3.3.2 PRINCIPLE OF SELECTION: EMPIRICAL FORMULA OR RADICAL LAW

Töpfer was one of the first to study the transformations of spatial representation in map form. He formulated the *principle of selection*, or *radical law* (Töpfer and Pillewizer, 1966), to express the relationship between map scale and the number of features represented on maps. Töpfer and Pillewizer found that many cartographic processes have a direct relationship between the square root of map scale ($\sqrt{S}$). Thus, the number (N) of cartographic features represented on a map is a function of $\sqrt{S}$, that is,

$$N = k \times \sqrt{S} \qquad (3.7)$$

Töpfer referred this formula to as the *radical law*. Substituting Equation 3.7 into Equation 3.6, the following equation can be obtained:

$$N_T = N_S \times \sqrt{\frac{S_S}{S_T}} \qquad (3.8)$$

Töpfer also called this formula the *law of natural dimension*. Variations can be made to Equation 3.8 by considering a number of factors such as the purpose of the map and the form of symbols, as follows:

$$N_T = C_P \times C_F \times N_S \times \sqrt{\frac{S_S}{S_T}} \qquad (3.9)$$

where C_F and C_P are the factors for map purpose and symbol form, respectively. C_F and C_P are also related to S_S and S_T. In the end, a more general formula is obtained as follows:

$$N_T = N_S \times \sqrt{\left(\frac{S_S}{S_T}\right)^x} \qquad (3.10)$$

where x takes a value of 0, 1, 2, 3, 4, or 5. This is called the *general selection law*. The interpretation of these values is as follows:

$$x = \begin{cases} 0 \Longrightarrow \textit{no selection} \\ 1 \sim 3 \Longrightarrow \textit{a densification of map image} \\ 4 \Longrightarrow \textit{equal map density} \\ 5 \Longrightarrow \textit{a loosening-up of map image} \end{cases} \qquad (3.11)$$

The principle of selection is able "to provide some measure of the amount of information which the cartographer can reasonably expect to put on a derived map," as emphasized by Maling in his introduction of Töpfer's paper (Töpfer and Pillewizer, 1966).

3.3.3 FRACTAL EXTENSION OF THE PRINCIPLE OF SELECTION

Equation 3.11 can also be written as

$$N_T = N_S \times \left(\frac{S_S}{S_T}\right)^p \tag{3.12}$$

where $p = x/2$ takes a value of 0, 0.5, 1, 1.5, 2, or 2.5. Yu (1993) found that the highest p-value belongs to areal symbols, then point symbols, then line symbols, and the least value was lettering. This means that the space occupied by areal symbols will decrease very quickly with a reduction in the map scale. In other words, the density of areal symbols will decrease quickly.

Equation 3.12 can be written as

$$p = \frac{x}{2} = \frac{\log(N_T/N_S)}{\log(S_S/S_T)} \tag{3.13}$$

Yu (1993) tried to connect the principle of selection with fractal dimension. In a similar use as the radical laws, if one knows the fractal dimension (D) of a line and its length (L_S) on the source map at scale 1:S_S, then it is possible to predict the length (L_T) of this line on the target map at scale 1:S_T. To do so, Equation 2.8 can be rewritten as

$$D = \frac{\log(L_T/L_S)}{\log(S_S/S_T)} \tag{3.14}$$

Then the following equation can be obtained for prediction of the line length at the target scale:

$$L_T = L_S \times \left(\frac{S_S}{S_T}\right)^D \tag{3.15}$$

Yu (1993) claims that this formula is a correlative form to Equation 3.12 and that Equation 3.14 is a correlative form of Equation 3.13. Yu further emphasizes that "This connection between Töpfer's 'Law' and fractal geometry is not simply a conversion. The theoretical and practical meaning goes beyond the simple mathematical expression and opens potential for future generalization formulation."

3.4   STRATEGIES FOR TRANSFORMATIONS OF SPATIAL REPRESENTATIONS IN SCALE

3.4.1   SEPARATION OF SCALE-DRIVEN FROM GRAPHICS-DRIVEN TRANSFORMATIONS

Map data is an important type of spatial representation. Paper maps have been used as a medium for both data storage and data display. As a result, in traditional manual map generalization, the transformations are carried out simultaneously for both the change of map complexity and the consideration of graphic legibility. In addition, the "characteristics and importance" of features are also considered in this process, as pointed out by Keates (1989). That is, one may want to keep certain characteristics on a map or to retain certain small features although they are too small for the map scale. All these together make the generalization process appear to be very subjective.

In a digital environment, graphic display and data storage are separated. Therefore, these two issues can be tackled separately. The legibility issue may be considered only when there is a need of graphic display because data resolution could be very high in a database. For example, two lines with spacing much less than 0.01 mm are still separable in a digital database. If the spatial data are only for analytical analysis, no legibility issue needs to be considered. Only when a graphic presentation is considered do we have the issues of graphic legibility, resulting in exaggeration, displacement, and other complex operations.

Müller et al. (1995) emphasized that a generalization can be separated into two stages: *model generalization* and *cartographic generalization*. A similar view has also been expressed by Li and Su (1995), as shown in Figure 3.8. In this figure the two stages are called *digital-to-digital transformation* (or *data generalization*) and *digital-to-graphic transformation* (or *graphic presentation*). There is a slight difference between these two views. In the one by Li and Su (1995), digital-to-graphics transformation is simply a graphic presentation but not a generalization process. Peng et al. (1996) employ a slightly different terminology, *database generalization* and *visualization generalization*, to express exactly the same view as Müller et al. (1995). All in all, there are two processes — one for data (or model) and the other for graphics.

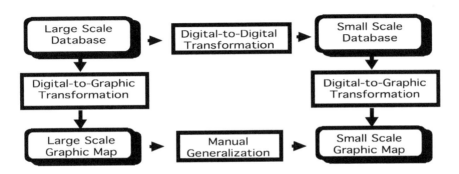

FIGURE 3.8 A strategy for digital map generalization (Reprinted from Li and Su, 1995. With permission.).

It can be noted here that the digital-to-digital transformation is the only step required if no graphic presentation is concerned. The digital-to-digital transformation is driven by scale. Such a process will simplify the shape, form, and structure of spatial representations and should be very objective, so that a unique solution can be achieved, given the same conditions. Such a transformation can be considered a transformation in scale dimension (Li, 1996) and it follows a natural principle. However, if graphics are considered, one needs to take into account the geographical, multipurpose, and cartographic requirements. It is now clear that cartographic requirements should be considered in the digital-to-graphic transformation after the scale-driven digital-to-digital transformation. Of course, one can also use some of the cartographic requirements as constraints for the digital-to-digital transformation. Some of the multipurpose and geographical requirements may also be used as constraints for this scale-driven transformation and for selecting data layers for generalization.

In the context of this book, emphasis is given to the scale-driven objective transformations. Indeed, only Chapter 10 is devoted to the transformations of graphic representation or, more precisely, displacement.

3.4.2 Separation of Geometric Transformation from High-Level Constraints

A map contains the following types of information about map features (Li and Huang, 2002):

- (Geo)metric information related to position, size and shape.
- Thematic information related to the types and importance of features.
- Relational information about spatial relations between neighboring features implied by distribution.

The transformations of spatial representation in the context of this book are about the geometric information, which is at the bottom level. Geometric transformations are achieved by some operations, each of which is implemented by one or more algorithms and operators. These operators and algorithms, such as affine and conformal transformations, are the basic functions in the transformation and can be utilized whenever needed.

The question of "when needed" should be answered by rules that are formalized by using thematic information and other knowledge. It has been recognized that such thematic information and knowledge can be acquired from (a) cartographic experts through interview, (b) existing maps through analysis, or (c) map specifications. A lot of work on this topic has been undertaken (e.g., Buttenfield and McMaster, 1991). However, this is a topic beyond the scope of this book.

After the geometric transformations are applied, the relations (order, topologic, and directional) between map features may be altered. The adequacy and allowable changes can be monitored by models for spatial relations. When human cartographers carry out generalization processing, they have an overview of a larger area and try to consider the interrelationship between features and thus consider a number of generalization operations simultaneously. However, computers do not have such an overview and execute operations one by one. Therefore, the relations between features need to be modeled so that reasoning changes before and after generalization can be made.

The thematic transformation is also directly related to topological relations between features, which vary greatly with scale. For example, on a map at 1:1,000 scale (i.e., a high degree of detail), almost every building and street is represented, and therefore, topological relationships between buildings and streets are important at this scale. However, on a map at 1:100,000 scale (a higher degree of abstraction), buildings need to be grouped together into blocks, and individual streets may disappear. In this case, the classes of features such as streets and buildings disappear and are replaced by new classes such as blocks. Therefore, topological relationships between blocks are important at this scale, and topological relationships between buildings and streets can and should be neglected. If the map scale is even smaller (a higher level of abstraction), then a town may become a point symbol, and thus all topological relations between features in the town disappear.

3.4.3 Distinguishing Three Levels of Transformations for Spatial Representation

The transformations of spatial representations can be carried out at three different levels:

- Individual features (i.e., feature level)
- A class of features (i.e., class level)
- The whole representation (i.e., map level)

At the feature level, one is concerned with the transformation of a specific map feature from source maps at a larger scale than the target maps at a smaller scale. A typical example is line generalization algorithms used to simplify a line to suit the representation at a smaller scale.

At the class level, one is concerned with the transformation of a specific class (or subclass) of features from the source maps at a larger scale to the target maps at a smaller scale. Many operators have been designed for transformations at this level, such as aggregation, merging, and typification.

At the map level, one considers the transformation of all classes of features from source maps at a larger scale to the target maps at a smaller scale as a whole. Map information is dealt with at this level. That is, one is concerned with the transformation of map information from larger scale source maps to smaller scale target maps (Knöpfli, 1983; Li and Huang, 2001). However, a discussion of transformations at this level lies outside the scope of this book.

In this book emphasis is given to transformations at feature and class levels. The discussions of transformations at the feature level are in Chapters 5–7, 9 and part of Chapter 11, while the discussions on the transformations at class level are in Chapter 8 and 10, and part of Chapter 11.

3.4.4 INTEGRATION OF RASTER-BASED MANIPULATION INTO VECTOR-BASED DATA STRUCTURE

Map generalization is due to the reduction in map space on smaller-scale maps. Therefore, the raster data model is the most appropriate one, as raster is a space-primary data structure. However, vector data is more intelligent. As a result, a hybrid data structure of vector and raster should be used as part of a strategy for multi-scale spatial representation. The vector structure could be used to hold spatial data in a database since a vector is a feature-primary data structure and is good for organizing data efficiently. The raster structure could be used as a working environment. This has been suggested by many researchers (e.g., Su et al., 1998; Peter and Weibel, 1999).

In this way, one needs to rasterize different types of vector features into various layers only when they need to be considered. This means it is not necessary to rasterize features if they are not relevant to a particular operation or if the original data are already in raster format. The raster size will be determined in such a way that rasterization will not affect the quality of the representation at the target scale. This can be achieved by following the natural principle, as presented in the previous subsection. After generalization is completed, the result can be vectorized back.

With such an integration in mind, a mix of vector-based and raster-based algorithms are presented in this book without special notifications.

REFERENCES

Buttenfield B. P. and McMaster, R. B., Eds., *Map Generalization: Making Rules for Knowledge Representation*, Longman Scientific & Technical, Essex, UK, 1991.

Gold, C. M. and Snoeyink, J., A one-step crust and skeleton extraction algorithms, *Algorithmica,* 30, 144–163.

Goodchild, M. and Quattrochi, D., Introduction: scale, multiscaling, remote sensing and GIS, in *Scale in Remote Sensing and GIS,* Quattrochi, D. and Goodchild, M., Eds., CRC Press, Boca Raton, FL, 1997, pp. 1–12.

Keates, J., *Cartographic Design and Production*, 2nd ed., Longman Scientific & Technical, Essex, UK, 1989.

Knöpfli, R., Communication theory and generalization, in *Graphic Communication and Design in Contemporary Cartography*, Taylor, D. R. F., Ed., John Wiley & Sons, 1983, pp. 177–218.

Li, Z. L., Transformation of spatial representation in scale dimension: a new paradigm for digital generalisation of spatial data, *International Archives for Photogrammetry and Remote Sensing,* XXI(B3), 453–458, 1996.

Li, Z. L. and Huang, P. Z., Transformations of spatial information in multi-scale representation, *Proceedings of the 20ᵗʰ International Cartographic Conference*, August 26–28, 2001, Beijing (CD-Rom).

Li, Z. L. and Huang P., Quantitative measures for spatial information of maps. *International Journal of Geographical Information Science*, 16(7), 699–709, 2002.

Li, Z. L. and Openshaw, S., Algorithms for automated line generalisation based on a natural principle of objective generalisation, *International Journal of Geographic Information Systems,* 6(5), 373–389, 1992.

Li, Z. L. and Openshaw, S., A natural principle for objective generalisation of digital map data, *Cartography and Geographic Information Systems*, 20(1), 19–29, 1993.

Li, Z. L. and Su, B., From phenomena to essence: envisioning the nature of digital map generalisation, *Cartographic Journal*, 32(1), 45–47, 1995.

Müller, J.-C., Weibel, R., Lagrange, J. P., and Salge, F., Generalization: state of the art and issues, in *GIS and Generalization,* Müller, J.-C., Lagrange J. P., and Weibel R., Eds., Taylor and Francis, London, 1995, pp. 3–17.

Peng, W., Tempfli, K., and Molennar, M., Automated generalization in a GIS context, in *Proceedings of Geoinformatics '96 (International Symposium on GIS/RS, Research, Development and Application)*, Florida, 1996, pp. 135–144.

Peter, B. and Weibel, R., Using vector and raster-based techniques in categorical map generalization, in *Third Workshop on Progress in Automated Map Generalization*, Ottawa, August 12–14, 1999.

Spiess, E., Map compilation, in *Basic Cartography*, Vol. 2, Anson, R. W., Ed., Elsevier, London, 1988, pp. 23–70.

Su, B., Li, Z. L., and Lodwick, G., Morphological models for the collapse of area features in digital map generalisation, *GeoInformatica,* 2(4), 359–382, 1998.

Töpfer, F. and Pillewizer, W., The principles of selection, *Cartographic Journal*, 3(1), 10–16, 1966.

Yu, Z., The Effects of Scale Change on Map Structure, PhD thesis, Department of Geography, Clark University, Worcester, MA, 1993.

4 Algorithms for Transformations of Point Features

In Chapters 2 and 3 some foundations were laid down. From this chapter on, various types of algorithms for multi-scale representation will be presented. This chapter will describe algorithms for point features.

4.1 ALGORITHMS FOR POINT FEATURES: AN OVERVIEW

On a two-dimensional (2-D) spatial representation, spatial features can be categorized into three types according to the geometric characteristics of symbols: point, linear, and polygonal (or areal). From this chapter on, the algorithms of various transformations for each type of these features, as mentioned in Chapter 1, will be presented sequentially.

Two of the transformations for individual point features, that is, elimination and exaggeration, are very simple and thus will be omitted in this text. The displacement of point features will be discussed in Chapter 11, together with the displacement of line and area features. In this chapter, only the algorithms for simplifying a set of point features will be described. Usually, a set of point features is called a *point cluster*.

Two types of spatial features can be regarded as point set (or *point cluster*) on a spatial representation. The first type includes the features that are represented by point symbols on the original (source) representation (maps), for example, control points, wells, fountains, small buildings, and spot heights. The second type includes the features that are represented by areal symbols on the original (source) representation but will be represented (or will be analyzed) as point features on target representations because of scale reduction, for example, a cluster of islands, lakes, and ponds. Figure 4.1 is a graphic illustration of the two types of point features. In this book we make no differentiation because our main concern is with the target map, that is, the representation of results after multi-scale transformations.

In summary, the algorithms to be discussed in the following sections are for aggregation, selective omission, structural simplification, and typification of a set of point features.

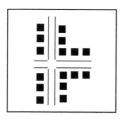

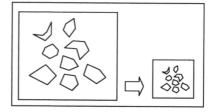

(a) Point features on source map   (b) Point features on target map

FIGURE 4.1 Two types of point features on a spatial representation.

4.2 ALGORITHMS FOR AGGREGATION OF A SET OF POINT FEATURES

As defined in Section 1.4.1, *aggregation of point features* means the categorization of a set of points into groups (classes or clusters) and then representation of each group by a single point. Therefore, this is a two-step process, and the critical part is the first step, which is often referred to as spatial clustering.

Spatial clustering is putting similar features into the same cluster (class or group). All *clustering algorithms* aim to minimize a measure of dispersion (in similarity) within the clusters. The measure of dispersion can be any of the following:

- Maximum distance to the cluster center (centroid) for any feature.
- Sum of the average distance to the center (centroid) over all clusters.
- Sum of the variance over all clusters.
- Total distance between all features and their center (centroid).

The similarity in nature could be measured by spatial, temporal, or socio-economic variables. In this context, distance is used as a general term to measure the similarity between two point features. That is, two close points will be classified in the same class.

Clustering techniques have been widely used for unsupervised classification of remote sensing images and spatial analysis. Clustering can be achieved sequentially or interactively. Clustering algorithms can be hierarchical or nonhierarchical. Therefore, there are many sequential hierarchical, sequential nonhierarchical, iterative hierarchical, and iterative nonhierarchical algorithms. In this chapter, only two commonly used algorithms, the K-means algorithm and the ISODATA (iterative self-organizing data analysis technique algorithm), are introduced.

4.2.1 K-MEANS CLUSTERING ALGORITHM

The basic idea of *K-means clustering* (MacQueen, 1967) is to partition (or classify) a set of N points into K clusters (groups or classes) that are mutually exclusive. The basic procedure is as follows:

1. Determine the number of clusters (K).
2. Designate a point as the cluster center for each of the K clusters.

3. Assign each of the N points to the nearest cluster (i.e., with shortest distance to the center, or centroid, of the cluster).
4. Compute the new center (centroid) for each of the K clusters after all points are assigned.
5. Repeat steps c and d until the centers (centroids) no longer move around significantly.

Figure 4.2 illustrates the process of K-means clustering. In this figure, 14 points are to be classified into 3 clusters. Points $K_{1,1}$, $K_{1,2}$, and $K_{1,3}$ are selected as the initial centers of these clusters (Figure 4.2a). After the first round of classification, six points are classified into cluster 1, five points into cluster 2, and three into cluster 3 (Figure 4.2b). The new centers are then computed and a second classification is carried out (Figure 4.2c). From the result of second classification, new centers are computed (Figure 4.2d). At this point the new centers do not move significantly, so

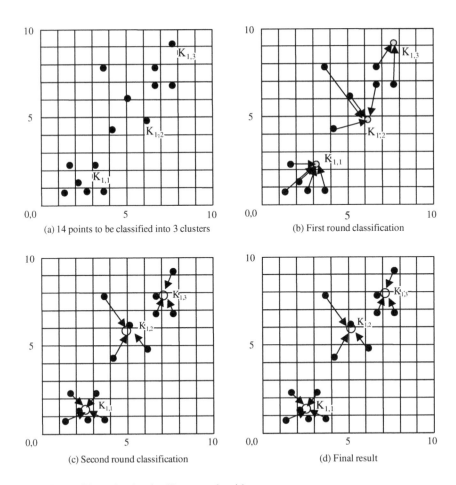

(a) 14 points to be classified into 3 clusters

(b) First round classification

(c) Second round classification

(d) Final result

FIGURE 4.2 Clustering by the K-means algorithm.

the classification procedure stops. In the second round, the K-means algorithm ensures that

$$d(P_{i,j}, C_i) \le d(P_{i,j}, C_l) \quad (i \ne l) \tag{4.1}$$

where P_{ij} is the jth point in cluster i; C_i is the center (or mean) of cluster i; C_l is the center (or mean) of cluster l; $d(P_{i,j}, C_i)$ is the distance from point $P_{i,j}$ to the center of cluster i.

4.2.2 ITERATIVE SELF-ORGANIZING DATA ANALYSIS TECHNIQUE ALGORITHM (ISODATA)

ISODATA (Tou and Gonzalez, 1974), as the name implies, is an interactive algorithm. It is not necessary to specify the number of clusters. It starts with a single cluster and applies a split-and-merge technique to progressively partition the points into more clusters through constantly assessing the similarity within a cluster (class or group). The similarity of points within a cluster is measured by the standard deviations of points in both the X and Y directions, that is, σ_x and σ_y. The procedure works as follows:

1. Determine the allowable values for the standard deviations, that is, $\sigma_{x,\max}$ and $\sigma_{y,\max}$.
2. Determine the number of clusters (K) and the number of iterations (n) (optional).
3. Treat all points as being in the same cluster to compute the means ($C_{old,X}$ and $C_{old,X}$) and the standard deviations (σ_x and σ_x) in both X and Y.
4. Determine whether there is a need to split the cluster. If $\sigma_x < \sigma_{x,\max}$ and $\sigma_x < \sigma_{y,\max}$, then stop splitting. If the specified number of iterations or number of clusters is reached, stop splitting. Then, if $\sigma_x > \sigma_y$, consider the X direction, or else consider the Y direction.
5. Split the cluster into two in the X direction if $\sigma_x > \sigma_y$ and $\sigma_x > \sigma_{x,\max}$. The temporary new centers are ($C_{old,X} - \sigma_x$) and ($C_{old,X} + \sigma_x$). Classify the points in the old cluster into two new clusters based on distance criterion. However,, if $\sigma_Y > \sigma_x$ and $\sigma_Y > \sigma_{y,\max}$, then the split will be in the Y direction.
6. For each of the new clusters, repeat steps 4 to 5.
7. Check each point to see whether the distance to its cluster centroid is the smallest among the distances to all centroids. If not, reclassify the point and recompute the corresponding centroids.

Figure 4.3 illustrates the process of ISODATA clustering. The 14 points shown in Figure 4.2 are used again. In this example, the number of clusters is defined as four, the number of iterations as three, and $\sigma_{x,\max}$ and $\sigma_{Y,\max}$ as 1.4. Figure 4.3a shows the consideration of all points as being within the same cluster (class or group), with point K_1 as the center. The standard deviations of this cluster in X and Y are 2.1 and 3.0, respectively. These values are larger than the thresholds. The first split is carried out in the Y direction because $\sigma_Y > \sigma_X$. $K_{2,1}$ and $K_{2,2}$ are the two new cluster centers. Clustering is then carried out based on distance, as shown in Figure 4.3b.

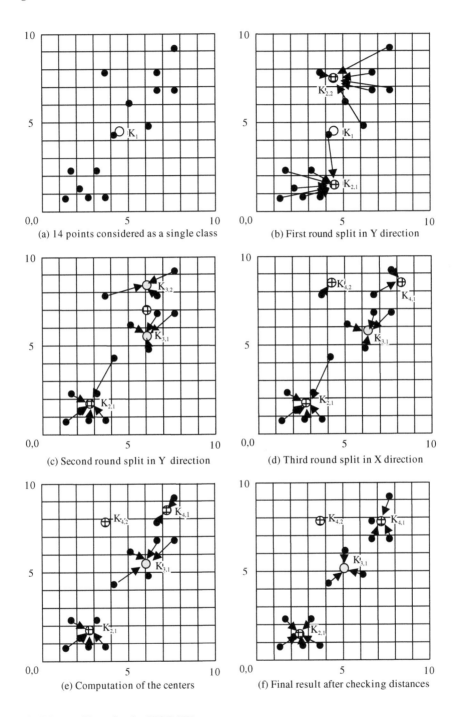

(a) 14 points considered as a single class

(b) First round split in Y direction

(c) Second round split in Y direction

(d) Third round split in X direction

(e) Computation of the centers

(f) Final result after checking distances

FIGURE 4.3 Clustering by ISODATA.

The centroids and the standard deviations of the two new clusters are then computed. It is found that the Y standard deviation of the cluster centered at $K_{2,2}$ is larger than $\sigma_{Y,max}$, and then a further split is carried out. The process is shown in Figure 4.3c. The centroids and standard deviations of these two new clusters are computed. It has been found that the X standard deviation of the cluster centered at $K_{3,2}$ is larger than $\sigma_{X,max}$, and then a further split is carried out, as shown in Figure 4.3d. The centroids of these two new clusters are then computed. In the end, each point is checked to see whether the distance to its cluster centroid is the smallest among the distances to all cluster centroids. It has been found that one point in the cluster centered at $K_{2,1}$ should be classified into the cluster centered at $K_{3,1}$. This point is then reclassified. As a result, the centroids of both clusters have been changed (Figure 4.3e). Consequently, a point in the cluster centered at $K_{3,1}$ is reclassified into $K_{4,1}$. Then the centroids of both clusters are moved, which causes a further point to be reclassified in the cluster centered at $K_{3,1}$ (Figure 4.3f).

4.2.3 DETERMINATION OF A REPRESENTATIVE POINT FOR A CLUSTER OF POINT FEATURES

After the first step of the aggregation operation, that is, clustering, all points are grouped into clusters. The next step is to represent the cluster by a single point. It can be imagined that the most representative point should be used. The question is, "What point is most representative?" From statistics, it can be found that the representatives are

- Mode: A point in an area with much higher distribution density compared with its surroundings (Figure 4.4a).
- Mean: The average coordinates in both X and Y coordinates (Figure 4.4b).
- Median: The point that partitions the planar space into two halves in X (i.e., left half and right half) and two halves in Y (lower and upper) to ensure equal numbers of points on each half in the same direction (Figure 4.4c).
- Nearest to mean: The point closest to the mean (Figure 4.4d).

It should be pointed out that the definition of *aggregation* given in this text may be different from others. Here, *aggregation* is referred to as the clustering followed by any of these four points (i.e., mode, mean, median, and nearest to the mean). However, the use of the mean to represent the cluster was termed as *typification* and the use of the point *closest to the mean as selection* by Jiang (2004).

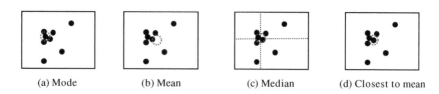

(a) Mode   (b) Mean   (c) Median   (d) Closest to mean

FIGURE 4.4 The most representative point of a point class cluster.

4.3 ALGORITHMS FOR SELECTIVE OMISSION OF A SET OF POINT FEATURES

Most *selective omission algorithms* are specially developed for selective omission of settlements. It is understandable that, when the scale of a map is reduced, not as many point symbols (e.g., settlements) can be represented. As a consequence, less important point features should be omitted. The importance of a point feature can be measured by one or more attributes. For example, a city may be signified by its population, gross domestic product, administrative status, or physical size. However, if some important point features are very close to an even more important feature, they may be omitted due to space problems while some less important points in sparse areas may be selected. For example, in eastern China large cities are located more closely together than in western China. When the map scale is reduced, some large cities in the east must be omitted because of a reduction in map space, while some medium cities in the west are retained. Figure 4.5 shows an example.

From the literature, it can be found that six algorithms have been proposed, five by Langran and Poiker (1986) (settlement-spacing ratio algorithm, distribution-coefficient algorithm, gravity-modeling algorithm, set-segmentation algorithm, and quadrant-reduction algorithm) and one by van Kreveld et al. (1995)

FIGURE 4.5 Selection and omission of cities as point features (http://china-hotelguide.com/).

(circle-growth algorithm). van Kreveld et al. pointed out that the set-segmentation and quadrat-reduction algorithms require a great deal of human intervention, so they are not suitable for automation. van Kreveld et al. also pointed out that the other three methods do not directly give a ranking of the base set, so changing the number of selected settlements involves recomputation. Indeed, the circle-growth algorithm is an improvement of the settlement-spacing ratio algorithm. Therefore, only the settlement-spacing ratio and circle-growth algorithms are described in this chapter.

4.3.1 SETTLEMENT-SPACING RATIO ALGORITHM

In the *settlement-spacing ratio algorithm,* a circle is used to indicate the significance of a point feature. The radius of the circle is inversely proportional to the importance of the point, that is,

$$R_i = \frac{C}{I_i} \tag{4.2}$$

where R_i is the radius of the circle for the ith point feature, I_i is the importance of the ith point feature, C is a constant, and $C > 0$.

In the selection process, points are tested in order of decreasing importance. That is, the most important point is tested first. A point will be selected only if its circle does not contain a previously selected point. In this way, important points close to a more important point may be omitted, while less important points with isolation may be selected. The critical step is the selection of an appropriate value for the constant C.

Figure 4.6 shows such a selection process. Figure 4.6a shows assets of five points with three levels of importance. Point A is the most important, point C is the least important, and points B, E, and D are the same level. Point A is selected first. A circle is then drawn from each point. It is found that the circle of point B contains point A, and thus point B is eliminated (Figure 4.6b). All the other points are retained, although point C is the least important (Figure 4.6c).

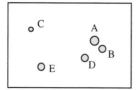   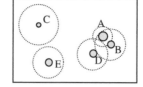   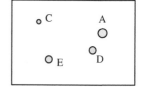

(a) A set of points   (b) Settlement-spacing ratio checking   (c) Result

FIGURE 4.6 Selective omission of point features by the settlement-spacing ratio algorithm.

4.3.2 CIRCLE-GROWTH ALGORITHM

The idea of the *circle-growth algorithm* is the opposite of the settlement-spacing ratio. In the former, a larger circle is drawn for the more important point feature. The radius of the circle is directly proportional to the importance of the point, that is,

$$R_i = C \times I_i \tag{4.3}$$

where R_i is the radius of the circle for the ith point feature, I_i is the importance of the ith point feature, C is a constant, and $C > 0$.

In the selection process the critical step is to rank each of the point features. To produce such a ranking, a circle is drawn from each point, with radius proportional to its importance, according to Equation 4.3. The initial value of C in Equation 4.3 is set such that no circles will overlap. Then the value of C is increased so that one or more circles of less important points will be contained by the circle of a more important point. The one covered by a larger circle is given a lower ranking, while the one with a larger circle is assigned a higher ranking. This process is repeated until the most important circle remains. In the end, points with low rankings will be deleted.

Figure 4.7 shows such a selection process by circle-growth algorithm. Figure 4.7a shows a set of five points with three different levels of importance. Figure 4.7b shows the first round of circle growth so that the circle of point A first covers that of point B. Point B is then given the lower ranking and is the first removed. Figure 4.7c shows the second round of circle growth so that the circle of point

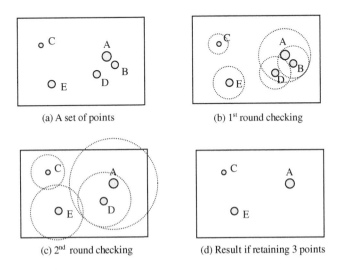

(a) A set of points   (b) 1ˢᵗ round checking

(c) 2ⁿᵈ round checking   (d) Result if retaining 3 points

FIGURE 4.7 Selective omission of point features by the circle-growth algorithm.

A first covers that of point D. Point D is then given the lower ranking and is removed if appropriate. This process continues until the last point remains. Figure 4.7d shows the result if one decided to select three points from the set of data.

The drawback of the circle-growth algorithm is that the points with very high importance have too much influence on the selection, and this results in the opposite of preserving density locally (van Kreveld et al., 1995). This is also true for other algorithms mentioned previously in this section.

4.4 ALGORITHMS FOR STRUCTURAL SIMPLIFICATION OF A SET OF POINT FEATURES

To simplify the structure of a set of point features for representation at a smaller scale, one needs to have a set of parameters for the description of a set of points (or point cluster). From the literature (e.g., Ahuja, 1982; Flewelling and Egenhofer, 1993; Guo, 1997; Yukio, 1997; Ai and Liu, 2002), it can be found that the following parameters are possible:

- Point number: The number of the point features in the set.
- Importance value: A value assigned to a point as a measure of its importance among point features.
- Voronoi neighbors: Points whose Voronoi regions are adjacent to that of a given point.
- Distribution range: One or more regions enclosing all point features of interest.
- Distribution density: The number of points in a unit area or the average distance between point features.
- Distribution modes: One or more areas with much higher distribution density compared with their surroundings.
- Distribution axes: One or more axes extracted from the area of a point cluster whose extent is linear.

Other parameters, such as shape, size, color, and orientation, have also been used (Yukio, 1997), but they concern the symbolization of point features.

4.4.1 Structural Simplification Based on Metric Information

In the literature, one may notice that there are not many algorithms for the *structural simplification* of a set of point features. This section describes the algorithm by Ai and Liu (2002).

Ai and Liu (2002) developed an algorithm for preserving the distribution characteristics of point clusters after simplification. They try to retain the density, centers, axes, and range of the distribution. Through analysis, they claimed that as long as the range is preserved, the axes are automatically preserved, and as long as the

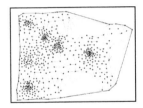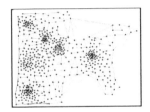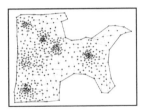

FIGURE 4.8 Defining the boundary of the point set by stripping the convex hull (Reprinted from Ai and Liu, 2002. With permission.).

relative density is preserved, the distribution centers are also preserved. Therefore, they tried to preserve the two more important parameters, that is, distribution range and relative density.

The working principle of this algorithm is as follows:

1. Determine the number of points to be retained based on the principle of selection discussed in Chapter 3 (Section 3.3.2).
2. Determine the range of the point distribution with two steps. The first step is to form a convex hull (see Chapter 8) of the points, and the second step is to strip the outer triangles, which is achieved by removing triangles whose outer edges are longer than a predefined value (Figure 4.8).
3. Simplify the boundary of the data points by a point-reduction algorithm (see Chapter 5) such as the Douglas–Peucker algorithm (Douglas and Peucker, 1973).
4. Compute a ranking for each point, taking value as $1/A_{VD,i}$, where $A_{VD,i}$ is the Voronoi region of the ith point.
5. Remove the point with the highest ranking from the data set. The procedure stops if the desired numbers are retained (or removed).
6. Update the new Voronoi diagram (and thus update ranking values) and consolidate the immediate neighbors (Voronoi neighbors); that is, suspend these points from the list.
7. Remove the point with the highest ranking from remaining point set. The procedure stops if the desired numbers are retained (or removed), or else repeat steps 6 and 7.
8. Release all the consolidated points for a second-round of selection. Repeat steps 5 and 7.

Figure 4.9 shows the structural simplification of a set of point features with the algorithm by Ai and Liu (2002). In this example 538 points were in the data set, and the distribution is shown in Figure 4.9a. The Voronoi diagram of this point set is shown in Figure 4.9b. After five rounds of selection, 265 points are retained, as shown in Figure 4.9c. The final result after graphic reduction is shown in Figure 4.9d.

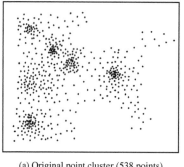

(a) Original point cluster (538 points)

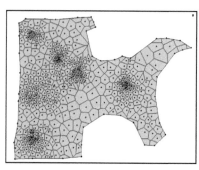

(b) Voronoi diagram of the point cluster

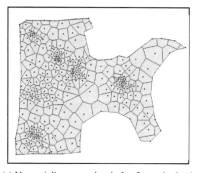

(c) Voronoi diagram updated after 5 round selection

(d) The result after simplification (265 points)

FIGURE 4.9 Structural simplification of a point set with the algorithm (Reprinted from Ai and Liu, 2002. With permission.).

4.4.2 STRUCTURAL SIMPLIFICATION CONCERNING METRIC AND THEMATIC INFORMATION

The algorithm by Ai and Liu (2002) concerns only the metric information of the points. However, as Li and Huang (2002) pointed out, four types of spatial information are contained in a cluster of spatial data (including point features):

- Statistical (positional).
- Metric.
- Thematic.
- Topologic.

That is to say, the algorithm by Ai and Liu (2002) is suitable for the structural simplification of a set of point features with the same thematic importance. However, in practice, some points are more important than others. In such a situation, Yan and Li (2004) suggest that the significance of each point be computed as follows:

$$S_i = I_i \times \frac{1}{A_{VD,i}} \tag{4.4}$$

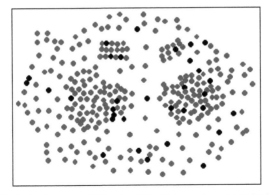

(a) Original point cluster (301 points, 32 with higher importance in black)

(b) Results from algorithm   (c) Results from algorithm
    by Ai and Liu (2004)   by Yan and Li (2005)

FIGURE 4.10 Comparison of results simplified by the algorithms with and without consideration of thematic information.

where S_i is the significance value of the ith point, I_i is the (thematic) importance value of the ith point, and $A_{VD,i}$ is the area of the Voronoi region of the ith point.

Figure 4.10 shows the simplification of a set of 301 points by the algorithms with and without the consideration of thematic importance. Figure 4.10a shows a dot map at 1:1:10,000 scale. Figure 4.10b shows the results simplified for representation at 1:50,000 scale without the consideration of thematic importance, and Figure 4.10c shows the result with the consideration of thematic importance. It is clear that more points with higher thematic importance are retained in Figure 4.10c.

4.5 ALGORITHMS FOR OUTLINING A SET OF POINT FEATURES: REGIONIZATION

Outlining a set of point features means grouping densely distributed point features into an area feature. McMaster and Shea (1992) referred to such an operation as *aggregation*. This is termed as a *regionization* operation in this text.

DeLucia and Black (1987) suggest a procedure as follows:

1. Generate the Voronoi diagram of the points.
2. Partition the Voronoi diagram into clusters with a given distance threshold.

3. Form Delaunay triangulation from the Voronoi regions.
4. Form the outline of each cluster by the boundary of the triangulation network.

However, the clustering algorithms described in Section 4.2 seem to be the more appropriate means for forming clusters from a set of point features. Therefore, the following algorithm is suggested:

1. Form clusters from the point set with a clustering algorithm.
2. Form a Delaunay triangulation network for each cluster
3. Strip the outer triangles by removing triangles whose outer edges are longer than a predefined value.
4. Form the outline of each cluster with the boundary of the triangulation network.

The *alpha-shape* proposed by Edelsbrunner and his collaborators (Edelsbrunner and Muker, 1994) can be used to cut off the triangular sides whose length is larger than the diameter of the alpha circle, so as to make clustering effects.

REFERENCES

Ahuja, N., Dot pattern processing using Voronoi neighbourhoods, *IEEE Transactions on Pattern Analysis and Machine Intelligence*, 4(3), 336–343, 1982.

Ai, T. and Liu, Y., A method of point cluster simplification with spatial distribution properties preserved, *Acta Geodaetica et Cartographica Sinica*, 25(1), 35–41, 2002. (in Chinese)

DeLucia, A. A. and Black, R. B., A comprehensive approach to automatic feature generalisation, *Proceedings of 13th International Cartographic Conference*, Morelia, Mexico, October 1987, 4, 169–192, 1987.

Douglas, D. H. and Peucker, T. K., Algorithms for the reduction of the number of points required to represent a digitised line or its caricature, *Canadian Cartographer*, 10(2), 112–122, 1973.

Edelsbrunner, H. and Muker, E. P., Three-dimensional alpha shapes, *ACM Transaction in Graphics,* 13, 43–72, 1994.

Flewelling, D. M. and Egenhofer, M. J., Formalizing importance: parameters for settlement selection from a geographic database, in *Proceedings of Auto-Carto 11 (11th International Conference on Automated Cartography)*, Minneapolis, 1993, pp. 167–175.

Guo, R. Z., *Spatial Analysis*, Wuhan Technical University of Surveying and Mapping Press, Wuhan, China, 1997. (in Chinese)

Jiang, B., Spatial clustering for mining knowledge in support of generalization processes in GIS, in *ICA Workshop on Generalization and Multiple Representation*, 20-21/08/2004, Leicester, 2004.

Langran, C. E. and Poiker, T. K., Integration of name selection and name placement, in *Proceedings of 2nd International Symposium on Spatial Data Handling*, 1986, pp. 50–64.

Li, Z. L. and Huang, P., Quantitative measures for spatial information of maps, *International Journal of Geographical Information Systems*, 16(7), 699–709, 2002.

MacQueen, J. B., Some methods for classification and analysis of multivariate observations, *Proceedings of 5th Berkeley Symposium on Mathematical Statistics and Probability,* Vol. 1, University of California Press, Berkeley, 1967 pp. 281–297.

McMaster, R. B. and Shea, K. S., *Generalization in Digital Cartography*, Association of American Geographers, Washington, DC, 1992.

van Kreveld, M., van Oostrum, R., and Snoeyink, J., Efficient settlement selection for interactive display, in *Proceedings of Auto Carto 12,* Bethesda, MD, 1995, pp. 287–296.

Tou, J. T. and González, R. C., *Pattern Recognition Principles,* Addison–Wesley Publishing Co., Reading, MA, 1974.

Yan, H. W and Li, Z. L., A Voronoi-based algorithm for point cluster generalization, in *Proceedings of 11th International Conference on Geometry and Graphics*, August 2004, Guangzhou, 2004. (CD-ROM)

Yukio, S., Cluster perception in the distribution of point objects, *Cartographica*, 34(1), 49–61, 1997.

5 Algorithms for Point-Reduction of Individual Line Features

In Chapter 4 some algorithms for the multi-scale representation of point features were presented. Naturally, the next topic is algorithms for line features. Four chapters are devoted to line features, as there is a huge body of literature in this area. This chapter discusses the point-reduction algorithms. The smoothing algorithms will be presented in Chapter 6, the scale-driven generalization algorithms in Chapter 7, and algorithms for a set of lines in Chapter 8.

5.1 ALGORITHMS FOR LINE POINT-REDUCTION: AN OVERVIEW

Point-reduction algorithms, as the name implies, aim to reduce the number of points required to represent a curved line. Reduction of the number of points on a line was an important issue in the early development of digital representation because of limited computing power. At that time, data volume was a big concern. As a consequence, many algorithms have been developed for such a purpose. Figure 5.1 illustrates the basic idea of point-reduction. In Figure 5.1a the original line is represented by many points, but only a few of them are selected by the algorithm to represent the original line. Figure 5.1b shows one of many possible results.

By removing some points, the shape of the line could be simplified in some cases, and thus many people have used such algorithms for the purpose of multi-scale representation (or generalization). Point-reduction algorithms try to make best approximation of the original line with a minimum number of points. It must be emphasized here that no scale change is involved in such an operation. That is, the output is for the representation of the line at the same scale. These algorithms are good for *weeding* operations. Therefore, it is misleading to use point-reduction algorithms for multi-scale representation (or generalization) purposes, as pointed out by Li (1993).

The idea behind the point-reduction algorithms is the psychological discovery by Attneave (1954) that some points on an object are richer in information than others and these few points (with richer information) are sufficient to characterize the shape of the object. In other words, a large number of points with less information can be removed without causing great deformation to the line. This also applies to spatial features.

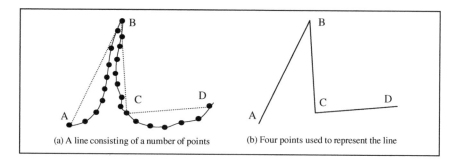

(a) A line consisting of a number of points   (b) Four points used to represent the line

FIGURE 5.1 The basic principle of point-reduction algorithms.

These points with richer information are referred to as *dominant points* in the computer graphics and pattern recognition literature and as *critical points* in geospatial science. Critical points of a curved line in classical geometry are:

- Maxima.
- Minima.
- Curvature minima.
- Curvature maxima.
- Points of infection.

To this list, Freeman (1978) added the following:

- Discontinuities in curvature.
- End points.
- Intersections (junctions).
- Points of tangency.

These points should be so well-defined that their characteristics will not be affected by scaling (enlargement or reduction), rotation, and translation.

Critical point is an important concept in many disciplines such as computer vision, image processing, pattern recognition, computer graphics, and geospatial science. For example, in computer vision and pattern recognition, the concept of critical points is used in algorithms for feature extraction, shape recognition, point-based motion estimation, and coding. In geospatial science it is used for data compression, for line caricature, and misleadingly for multi-scale representation of lines (Li, 1993).

Data points on a line can be reduced by two basic approaches:

- Corner detection: Directly identify points with greater significance and then select them based on a given threshold.
- Curve approximation: Indirectly obtain points through the approximation of a curved line using a number of linear segments (as shown in Figure 5.1).

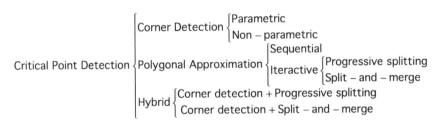

FIGURE 5.2 Classification of algorithms for detecting critical points (Reprinted from Li, 1995).

Curve approximation is also referred to as *polygonal approximation*. In this approach, another two solutions can be identified: sequential methods and iterative methods. Two types of implementations are commonly employed in iterative approaches: progressive splitting and split-and-merge. In corner detection, two commonly used approaches are parametric and nonparametric. There are certain advantages and disadvantages to both corner detection and polygonal approximation approaches. Therefore, the natural line of thought is to combine these two so that the advantages of both can be kept and shortcomings off-set. Hybrid algorithms might also be classified into two groups: corner detection followed by a split-and-merge process and corner detection followed by a progressive splitting process. Figure 5.2 shows such a classification by Li (1995).

From the literature, it can be found that there are many ways to classify point-reduction algorithms. Table 5.1 lists some examples.

Almost all algorithms appearing in the cartographic literature belong to the category of polygonal approximation, except for a few (e.g., Thapa, 1988b). A review of those algorithms developed before the mid-1980s have been provided by McMaster (1987) and Thapa (1988a).

TABLE 5.1
Classification of Point-Reduction Algorithms

Criteria	Types	Characteristics	Reference
Iteration	Direct	Obtain the critical points in one pass through the data	Thapa, 1988a
	Iterative	Need more than one pass through the data	
Area of processing	Local	Consider a few points at a time	McMaster, 1987
	Global	Take into consideration all the points	
Object of processing	Scan-along	Process the points one at a time	Pavlidis, 1978;
	Hop-along	Process segments of a line through split-and-merge	Thapa, 1988a

Many terms are in use for point-reduction algorithms, such as (Thapa, 1988a):

- Planer curve segmentation.
- Polygonal approximation.
- Vertex detection.
- Piece-wise linear approximation.
- Curve partitioning.
- Data compaction.
- Straight line approximation.
- Selection of main points.
- Detection of dominant points.
- Determination of main points.

Through a critical analysis, it can be found that the criteria used for polygonal (or curve) approximation are primary geometric parameters (such as distance, angle, and area), while the criteria used for corner detection are functions of some primary geometric parameters (such as the ratio between distance and chord and the cosine of an angle). Such functions are the estimates of curvature. In this chapter a new classification scheme is adopted that takes into consideration the following two parameters, that is, the criteria used in the algorithm and iteration of the processing. Under this scheme, this chapter covers:

- Sequential algorithms with geometric parameters as criteria.
- Iterative algorithms with geometric parameters as criteria.
- Algorithms with a function of geometric parameters as criteria.

5.2 SEQUENTIAL ALGORITHMS WITH GEOMETRIC PARAMETERS AS CRITERIA

The *sequential algorithms* start from a point to find the longest allowable segment within which all points along the original line can be ignored with a given criterion (tolerance or threshold). The last point on the longest allowable segment is considered as being a critical point and is designated as the next starting point. This process is repeated until the whole curve has been traced. At the end, all these detected points plus the two end points are considered as being the critical points of this curve line.

5.2.1 ALGORITHM BASED ON NUMBER OF POINTS

The simplest algorithm for reduction of points on a line is based on the number of points (see McMaster, 1987; Li, 1995), called *N*th points. This algorithm selects points with an interval of N points. Figure 5.3 illustrates the principle. In this figure there are 12 points. If $N = 3$, then every third point is selected, and points 1, 4, 7, 10, and 12 will be selected in the end. If $N = 4$, then every fourth point will be selected. As a result, points 1, 5, 9, and 12 will be selected.

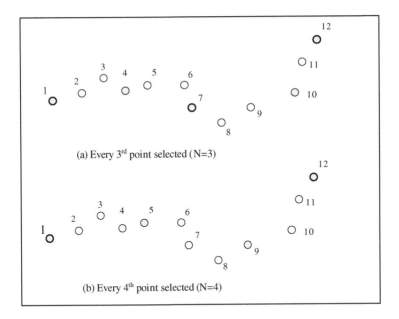

(a) Every 3rd point selected (N=3)

(b) Every 4th point selected (N=4)

FIGURE 5.3 Selection of points by a simple Nth point algorithm.

It is obvious that this algorithm is very simple and does not consider the richness of information content of the points at all. Thus, the points retained (or selected) may not be of any particular importance. Theoretically speaking, this is not acceptable. However, in practice, if the points on a line are very dense, such as in the case of using stream-mode digitization with a short distance as the criterion, this simple algorithm works very well.

5.2.2 Algorithm Based on Length

Another simple algorithm is the selection of a point at a certain distance from the previously selected point (see McMaster, 1987; Li, 1995), which is called length in this text to clearly differentiate this criterion from the perpendicular distance. Figure 5.4 illustrates these terminologies. In this figure, L is the length between points A and B, while D is the perpendicular distance from B to line A-C.

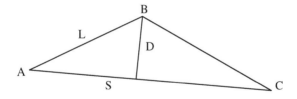

FIGURE 5.4 Two basic parameters: length and perpendicular distance.

The basic idea is that the variations between two points may be ignored if the distance (length) between them is smaller than the given criterion. The algorithm works as follows:

1. Select the two end points.
2. Compute the distance (length) between the current point and the previously selected point.
3. Select the current point if the length is larger than the given criterion, or else consider the next point.
4. Repeat steps 2 and 3.

Figure 5.5 illustrates the principle of such an algorithm. In Figure 5.5a the distance between the starting point and point 2 is smaller than the criterion given, but the distance to point 3 is larger than the criterion, so point 3 is selected. From point 3, the distances to points 4 and 5 are both smaller than the criterion, but the distance to point 6 is larger than the criterion, so point 6 is selected. The process continues, and in the end, points 1, 3, 6, 9, 11, and 12 are selected.

An improvement of this algorithm is to employ a cumulative length (instead of chord length) as the criterion. The cumulative length is the sum of all the lengths between two successive points at the interval from the previously selected point to the current point that is under consideration. The cumulative length is the arc length. Figure 5.5b illustrates such a variation in results. For example, when considering point 8, the sum of the length between points 6 and 7 and the length between 7 and 8 is considered. In this case the sum is larger than the criterion, so point 8 is selected.

Again, this algorithm is very simple and it does not consider the information contents of the points at all.

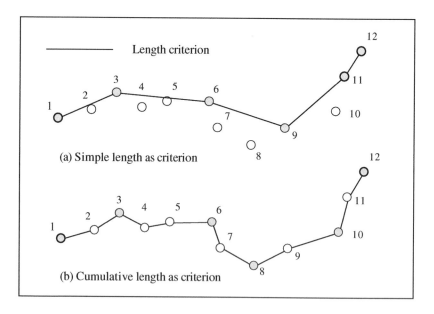

FIGURE 5.5 Selection of points with length as the criterion.

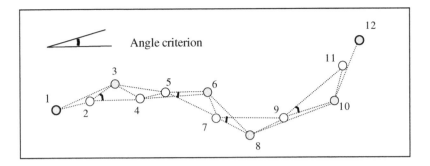

FIGURE 5.6 Selection of points by simple angle criterion.

5.2.3 ALGORITHM BASED ON ANGLE

Another simple algorithm is the use of angle as the criterion (see McMaster, 1987; Li, 1995). The basic idea is that points can be deleted if their removal causes a small angular distortion. The algorithm works as follows:

1. Select the two end points.
2. Compute the angle formed by the current point, the previous point, and the next point.
3. Select the current point if the angle is larger than the given criterion.
4. Continue the process point by point.

Figure 5.6 illustrates the principle. In this figure, the angle formed by points 2, 1, and 3 is first checked. If the angle is smaller than the given criterion, then point 2 is deleted. Next, the angle formed by points 3, 2, and 4 is checked. If it is larger than the criterion, point 3 is retained.

This algorithm seems better than the two described previously. However, it is not widely used in practice, possibly because (a) the computation of angles is not efficient and (b) a large shape distortion may be caused with cumulative deletion of points with small angles.

5.2.4 ALGORITHM BASED ON PERPENDICULAR DISTANCE

This algorithm was used in a drawing device, Geograph 4000, provided to the Experimental Cartography Unit of Britain by German forms AEG and Aristo. Thus, we refer to it as the *AEG algorithm*. This algorithm was reported by Lang (1969) and works as follows:

1. Select the two end points.
2. Compute the perpendicular distances from all the points between the current point and the previously selected point.
3. Consider the next point if none of such distances is larger than the criterion, or else select the one before the current point.
4. Repeat steps 2 and 3.

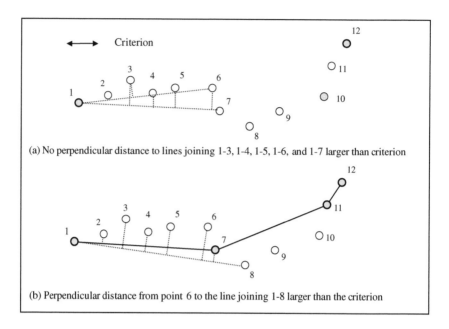

(a) No perpendicular distance to lines joining 1-3, 1-4, 1-5, 1-6, and 1-7 larger than criterion

(b) Perpendicular distance from point 6 to the line joining 1-8 larger than the criterion

FIGURE 5.7 The working principle of the AEG algorithm.

The AEG algorithm's working principle is illustrated in Figure 5.7. In this figure point 1 is the starting point of the line segment. From this point, a line can be drawn to connect point 3. The distance from point 2 to line 1-3 is then compared with the criterion given to see whether this distance is greater than the criterion. If is it, then point 2 (instead of point 3) will be selected as a critical point and used as next starting point. If it is not greater, then a line is drawn from point 1 to point 4, and the distances from point 2 and point 3 to this line are compared with the criterion to see whether any of them is greater than the criterion. If they are, point 3 (instead of point 4) will be selected as a critical point and used as next starting point. It is not true in this case, so this process continues until point 8 is tested. That is, the distance from point 6 to line 1-8 is greater than the criterion. Now point 7 is selected and used as the next starting point. The process continues with point 7 as the new starting point.

5.3 ITERATIVE ALGORITHMS WITH GEOMETRIC PARAMETERS AS CRITERIA

Sequential methods have the drawback of missing some important features. In the case of the example given in Figure 5.7, points 6 and 8 are missed although they are richer in information than point 7. Due to this serious drawback, the author does not know of much recent development of algorithms of this category. In this section some *iteractive algorithms* will be described.

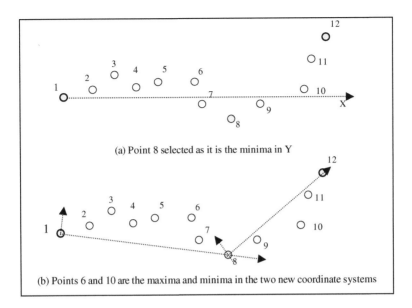

(a) Point 8 selected as it is the minima in Y

(b) Points 6 and 10 are the maxima and minima in the two new coordinate systems

FIGURE 5.8 Selection of points based on the minima and maxima of the X and Y directions using the Li algorithm.

5.3.1 ALGORITHM BASED ON MINIMA AND MAXIMA

One of the algorithms in this category, developed by Li (1988), makes use of the minima and maxima in both the X and Y directions. The *Li algorithm* works as follows:

1. Select the two end points.
2. Select the points that are minima and maxima in both the X and Y directions (Figure 5.8a).
3. The selected points will partition the curve into two segments. Transform points on each segment into a new coordinate system by a rotation, in which the two end points of the segment (Figure 5.8b) are used as the controls.
4. The process stops if the given criterion is met; otherwise, repeat steps 2 and 3.

The criterion could be the number of points selected, the rounds of selection, the coordinate values of the maxima and mixima, or a combination of two of these.

Figure 5.8 illustrates this principle. In the first round, the original coordinate system is used. Point 8 is the minima and point 12 is the maxima in Y, and point 1 is the minima in X and point 12 the maxima in X. Therefore, point 8 is selected in the first round, and the curved line is partitioned into two segments. Points 1 and 8 will used to perform a coordinate transformation (i.e., a rotation only) for the first segment, and point 6 is the maxima in Y in the new coordinate system. Similarly, point 10 is the local minima in the new coordinate system of second segment. Li (1988, 1993) has claimed that this algorithm works very efficiently.

5.3.2 PROGRESSIVE SPLITTING BASED ON PERPENDICULAR DISTANCE

The typical example of progressive splitting algorithms is the *Douglas–Peucker algorithm*. In this algorithm the curved line between two initial points is split into two segments by a partitioning point between them. This partitioning point is selected as a critical point and will become a new starting point for both of the two new curve segments, which will then be split again until a criterion is met. The algorithm works as follows:

1. Select the two end points as initial critical points.
2. Draw a line between two neighboring critical points and compute the perpendicular distances from all the points between the these two points to the drawn line.
3. Consider the point with the largest perpendicular distance. If this distance is larger than the criterion (threshold), select the point as a new critical point. This point partitions this curve segment into subsegments.
4. For each of the subsegments, repeat steps 2 and 3 unless no more points can be selected from any of the subsegments.

The working principle of this algorithm is illustrated in Figure 5.9. The algorithm starts with point 1 (the first point) and point 12 (the last point) to find out the greatest distances from all points in between these two points to line 1-12. That is the distance from point 8 in this example. This distance is then compared with the given criterion. If it is greater than the criterion, point 8 is selected as a critical

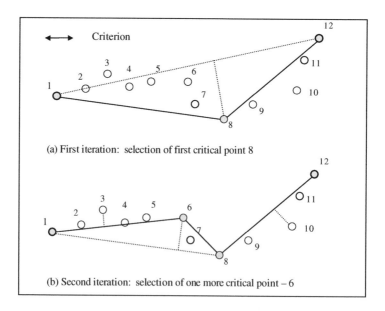

(a) First iteration: selection of first critical point 8

(b) Second iteration: selection of one more critical point – 6

FIGURE 5.9 The working principle of Douglas–Peucker algorithm (also known as the Ramer [1972] algorithm and the Forsten algorithm).

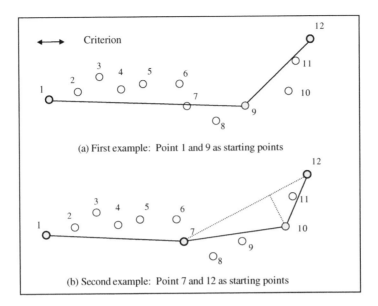

FIGURE 5.10 Variation of critical points detected by the Douglas–Peucker algorithm with two initial points.

point and it then splits the curve into two segments. It also becomes the new starting points for both of new curve segments. The first segment is again split into two by point 6, and the second by point 10. Then points 1, 6, 8, 10, and 12 are the critical points detected.

The critical points detected by this algorithm vary with the locations of the two initial critical points. For the example given in Figure 5.10, if one starts with point 1 and point 9, then all points in between these two points will not be detected even though point 8 is very important even for the curve segment from point 1 to point 9 only (see Figure 5.10). In this case points 1, 9, and 12 will be the critical points detected. This is one of the main drawbacks of this algorithm. Another shortcoming will be discussed in Section 5.6.

This algorithm was first published by Ramer in *Computer Graphics and Image Processing* in 1972. Therefore, it is known as the Ramer algorithm in computing science. A year later, the same algorithm was published in *The Canadian Cartographer* by Douglas and Peucker (1973), so it is known as the Douglas-Peucker algorithm in the geospatial science community. This algorithm has also been described by Duda and Hart (1973), who called it the Forsen algorithm.

5.3.3 SPLIT-AND-MERGE BASED ON PERPENDICULAR DISTANCE

Another iterative method is the so-called *split-and-merge*. There are many split-and-merge algorithms. In this section only the algorithm described by Ansari and Delp (1991) is described.

The *Ansari–Delp algorithm* works as follows:

1. Assign an arbitrary number of points along the line as initial critical points.
2. For each pair of critical points, the perpendicular distances from all points to the line joining these two critical points are computed. If any of the perpendicular distances is greater than the criterion, then the point with the largest perpendicular distance is selected as a critical point.
3. For each pair of adjacent line segments, the three consecutive critical points will be checked to see whether the middle critical point should be retained. The perpendicular distance from this point to the line is then compared with the threshold. If it is shorter than the criterion, then the middle point is removed from the list.
4. Repeat steps 2 and 3 until no further split-and-merge is required.

Figure 5.11 illustrate the working principle of this algorithm. Points 1, 4, 10, and 12 in Figure 5.11a are the initial critical points. For the pair consisting of points 4 and 10, the perpendicular distances of all points to the line joining these two critical points are computed. The perpendicular distance from point 8 is the largest and is larger than the criterion, so point 8 is selected as a critical point (Figure 5.11b). In other words, the curve segment between points 4 and 10 is split by point 8. The three consecutive critical points 1, 4, and 8 in Figure 5.11b are checked to see whether point 4 can be removed from the list of critical points. The perpendicular distance from point 4 to the line (i.e., the line joining points 1 and 8) is then compared with the threshold. The distance is shorter than the criterion, so point 4 is removed from the list and these two line segments are then merged as shown in Figure 5.11c. This merge process is also be applied to the segment with points 8, 10, and 12. In the end, points 1, 6, 8, and 12 are selected.

Again, the performance of the split-and-merge methods is very sensitive to the setting of the initial critical points partitioning the curve. For example, if points 1, 4, 9, and 12 are used as the initial points, then the results will be identical to those shown in Figure 5.10. That is, points 6 and 8 are not detected, and the final list of critical points will be points 1, 9, and 12.

This split-and-merge algorithm locates the critical points on the line itself. However, for other methods, the detected points may not lie on the original curve. An example of such algorithms is the one developed by Pavlidis and Horowitz (1974). In this algorithm, points on a curve are approximated by interpolating straight line segments.

5.3.4 Algorithm Based on Area

Visvalingham and Whyatt (1993) claimed that the size of an area "sets a perceptual limit on the significance" and is "the most reliable metric for elimination since it simultaneously considers distance between points and angular measures." They used the *effective area* of a point as the criterion for point reduction. The concept of effective area is illustrated in Figure 5.12a. In this figure all the effective areas for

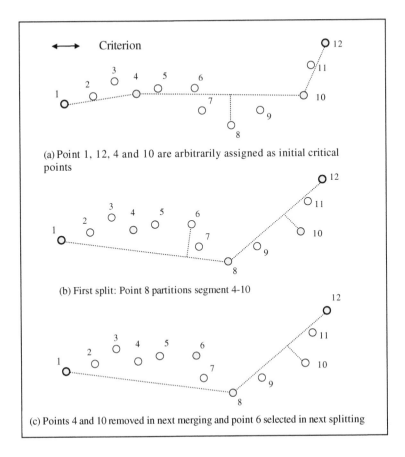

FIGURE 5.11 The working principle of the Ansari-Delp algorithm.

points 2 to 11 are shaded. For example, the effective area of point 2 is the area covered by the triangle formed by points 1, 2, and 3.

The basic idea of the *Visvalingham–Whyatt algorithm* is to progressively eliminate points with small effective areas so as to ensure the least area displacement. The procedure is as follows:

1. Compute the effective area of each point (Figure 5.12a).
2. Delete the point with the smallest effective area from the original list and add this point to the new list together with its associated area.
3. Update (recompute) the effective areas of the two points adjoining the point just deleted (Figure 5.12b).
4. Repeat steps 2 and 3 until a criterion is met (Figure 5.12c,d).

The criterion could be the size of the area or the number of points to be retained.

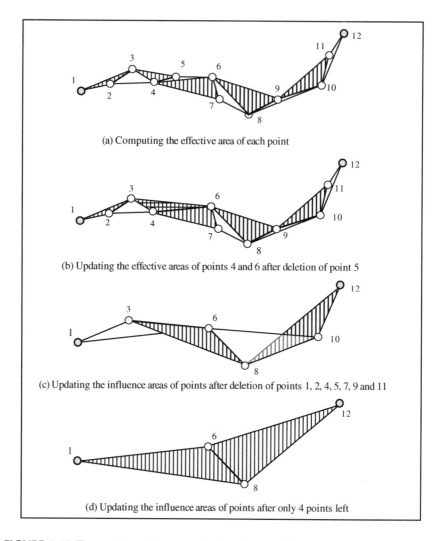

(a) Computing the effective area of each point

(b) Updating the effective areas of points 4 and 6 after deletion of point 5

(c) Updating the influence areas of points after deletion of points 1, 2, 4, 5, 7, 9 and 11

(d) Updating the influence areas of points after only 4 points left

FIGURE 5.12 The working principle of the Visvalingham–Whyatt algorithm.

5.4 ALGORITHMS WITH FUNCTIONS OF GEOMETRIC PARAMETERS AS CRITERIA

It is well-known that curvature is a parameter used to measure the importance of a point on a line. However, the calculation of curvature involves the first and second derivatives. Such a calculation is therefore considered computationally expensive. Therefore, alternatives are sought to represent the curvature values. These alternative measures are usually the functions of the basic geometric parameters. Some of these functions have been used as criteria in point-reduction algorithms, and such algorithms belong to the category of *corner detection*.

The common stages for corner detection are:

1. Estimate the curvature for each point on the curve.
2. Locate the points that have local maximum (both positive and negative) curvatures as the corners.

Step 2 can be subdivided into another two steps: (a) a threshold (whether input or computed during the process) is applied to the curvature estimates to eliminate points whose curvature is absolutely too low for them to be critical points, and (2) a process of nonmaxima suppression is applied to the remaining points to further eliminate any point whose curvature estimate is not local maxima in a sufficiently large segment (i.e., region-of-support) on the curve.

The alternative parameter for the curvature of a point P is estimated using a set of points centered at P. The collection of points at a certain distance from each side of P is known as the *region-of-support*. In the development of algorithms used for corner detection, the determination of the region-of-support is an important issue, apart from the selection of curvature estimate.

5.4.1 ALGORITHM BASED ON COSINE VALUE

In a classic algorithm developed by Rosenfeld and Johnston (1973), the cosine value of a point, $\cos(\theta_{i,k})$, is used to approximate the curvature of the point. Here, $\theta_{i,k}$ is the angle between the line joining point I and point $(i - k)$ and the line joining point I and point $(i + k)$, where $k = 1, 2, \ldots m$. In this algorithm, m is a smoothing factor and is an input parameter. A larger m makes the results smoother. In the case shown in Figure 5.13, $i = 6$ and $k = 2$. The value of $\cos(\theta_{i,k})$ is termed a k-cosine angle measure and lies in the range between (-1) (for a straight line) and $(+1)$ (sharpest angle; i.e., 0 degrees).

The Rosenfeld–Johnston algorithm works as follows:

1. Determine a value for m.
2. For each point I, compute all the $\cos(\theta_{i,k}) \cos(\theta_{i,k})$ values with k varying from m to 1.
3. $K = k$ is determined to be the region-of-support for point I, if $k = 1$ or $\cos(\theta_{i,k+1}) < \cos(\theta_{i,k}) > \cos(\theta_{i,k-1})$ for the first time.
4. Point I will be retained as critical point only if $\cos(\theta_{i,k}) \geq \cos(\theta_{j,k})$ for all j within the range $\left[\left(i - \frac{K}{2}\right), \left(i + \frac{K}{2}\right)\right]$.

Figure 5.13 illustrates the working principle of this algorithm. Assume that $m = 4$. Let us consider point 8. From Figure 5.13b, it can be found that when $k = 3$, $\cos(\theta_{8,4}) < \cos(\theta_{8,3}) > \cos(\theta_{8,2})$; thus, $K = 3$. Then points 6, 7, 9, and 10 will be considered. It can be found from Figure 5.13c that $con(\theta_{8,3}) > con(\theta_{6,3})$, $\cos(\theta_{8,3}) > \cos(\theta_{7,3})$, $\cos(\theta_{8,3}) > \cos(\theta_{9,3})$. Therefore, point 8 is retained as the critical point.

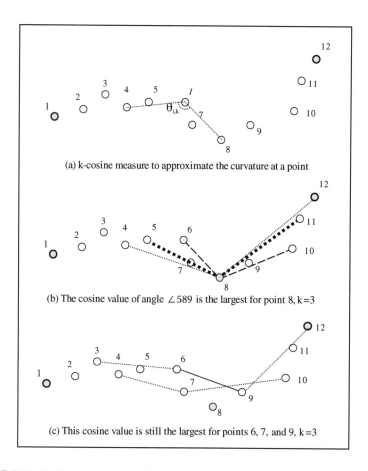

(a) k-cosine measure to approximate the curvature at a point

(b) The cosine value of angle ∠589 is the largest for point 8, k=3

(c) This cosine value is still the largest for points 6, 7, and 9, k=3

FIGURE 5.13 Working principle of the Rosenfeld–Johnston algorithm.

5.4.2 Algorithm Based on Distance/Chord Ratio

In the Rosenfeld–Johnston algorithm, the m value is an input parameter for the determination of the region-of-support. Methods with input parameters are sometimes called parametric methods. Two problems are associated with such methods. The first is that the line is assumed to be unknown, so it is difficult to determine an appropriate parameter without trial and error. The second problem is that the computation of local curvature measures based on this single region-of-support is ineffective for lines with varying level of detail. To overcome this problem, an adaptive approach to determine the region-of-support for each point on a curve without using any input parameter is widely used. The algorithm developed by Teh and Chin (1989) belongs to this category.

In the *Teh–Chin algorithm,* instead of the cosine values, the ratio (d/S) between the distance (d) from point i to the chord joining points ($i - k$) and ($i + k$) and the length of the chord itself (S) is used to determine the region-of-support for point i,

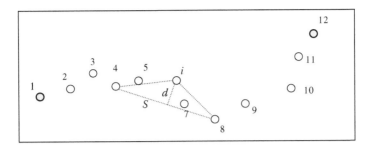

FIGURE 5.14 The Teh a–d Chin algorithm, using *d/S* to determine the region-of-support.

as shown in Figure 5.14. To do so, starting from $k = 1$, compute the distance $d_{i,k}$ and chord $S_{i,k}$, until one of the following conditions is met:

$$S_{i,k} \geq S_{i,k+1} \tag{5.1}$$

or

$$\begin{cases} \dfrac{d_{i,k}}{S_{i,k}} \geq \dfrac{d_{i,k+1}}{S_{i,k+1}}, & \text{if } d_{i,k} > 0 \\[3mm] \dfrac{d_{i,k}}{S_{i,k}} \leq \dfrac{d_{i,k+1}}{S_{i,k+1}}, & \text{if } d_{i,k} < 0 \end{cases} \tag{5.2}$$

Then a measure of significance, for example, the *k*-cosine measure by Rosenfeld and Johnston (1973) or the *k*-curvature measure (i.e., the true curvature computed by using the same three points as used for the *k*-cosine measure), is computed for each point using the points within its region-of-support. Finally, nonmaxima in curvature are suppressed in the process, similar to that in the Rosenfeld–Johnston algorithm. The remaining points are considered to be critical points.

This algorithm does not require any input parameter and works well on a curved line that is not corrupted with noise. However, it is sensitive to noise in the digital curve, and false critical points will be detected because of noise (Ansari and Huang, 1991). In addition, it is computationally expensive since a region-of-support needs to be determined for each point.

5.4.3 Algorithm Based on Local Length Ratio

Instead of the ratio (*d/L*) between the perpendicular distance (*d*) and the length of chord used by Teh and Chin (1989), the ratio between the arc length between two points and the corresponding chord length is used as the criterion by Nako and Mitropoulos (2003). This ratio,

$$LLR = \frac{L}{S} \tag{5.3}$$

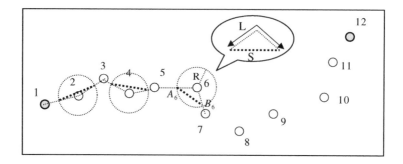

FIGURE 5.15 The Nakos–Mitropoulos algorithm, using LLR as the criterion.

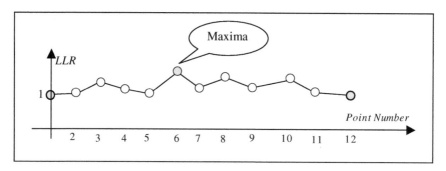

FIGURE 5.16 The Nakos–Mitropoulos algorithm, using maxima of LLR as the critical point.

is called the *local length ratio* (LLR) and is illustrated in Figure 5.15. Taking point 6 as an example, the arc length is the sum of two segments, that is, the distance between point A_6 and point 6 plus the distance between point B_6 and point 6. Here A_6 is the intersection between the circle and the line segment connecting points 5 and 6. The circle is centered at point 6 and has a fixed radius R. B_6 is the intersection between the circle and the line segment connecting points 6 and 7.

The Nako–Mitropoulos algorithm works as follows:

1. Determine a value for the radius of the circle R.
2. Compute an LLR value (LLR_i) for each point.
3. Construct a function for $LLR_i = f(i) = L_1/S_1$ and locate the maxima of the function (see Figure 5.16).
4. Take the maxima as critical points.

5.5 EVALUATION OF POINT-REDUCTION ALGORITHMS

The previous sections describe a number of algorithms for the reduction of points on curved lines. These typical examples were included because there are too many of this type of algorithm to be included in this chapter.

A question arising here is, "How well do they perform?" or "How do we evaluate the results?" Only after this question is answered, can the algorithms be improved. This section is devoted to this evaluation issue.

5.5.1 MEASURES FOR EVALUATION OF POINT-REDUCTION ALGORITHMS

From the literature, it can be summarized that the following questions need to be answered for the assessment of point-reduction algorithms:

1. Are the points detected really critical to the representation of the line?
2. Are the points detected accurately located on the line?
3. Is the line formed by the retained points deviated too much from the original line?
4. Is there any topological conflict with the new lines formed by the retained points.

The first question can be answered (Visvalingham and Whyatt, 1990):

1. Visually inspecting of the results.
2. Comparing the results obtained from different data sets, which are the consequences of performing different types of geometric transformation, that is, scaling, rotation, and translation.
3. Comparing the results obtained from different sets of initial points.

Related to the second question, there are two types of inaccuracy. The first type is caused by the smoothing of original line. That is, the critical point detection is performed on the smoothed line in order to minimize the noise effect. As a result, some points on the smoothed line are selected as critical points. There is certainly a deviation between the point on the smoothed line and the corresponding point on the original line. The second type of inaccuracy is the location of wrong point on the line, possibly owing to the noise. This is not a serious problem here because the algorithms described in this text work on the original lines, which are not noise-prone.

To answer the third question, White (1985) and McMaster (1986) have developed set of measures. The vector and areal displacements are in common use. These two parameters are illustrated by Figure 5.17. The areas formed by the original line and

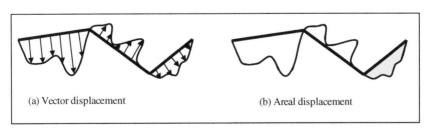

(a) Vector displacement   (b) Areal displacement

FIGURE 5.17 Vector and areal displacements for assessment of the quality of point-reduction.

the new line (formed by retained points) are called sliver polygons. Of course, the smaller the vector displacement and areal displacement, the better the algorithm performs. From the vector displacement, the largest error, median, and mean errors can be computed. The area of the sliver polygon can be divided by the length of the new line to produce a unit displacement.

To answer the last question, a number of experimental evaluations have been conducted (e.g., Visvalingham and Whyatt, 1990; Li and Openshaw, 1992; Weibel, 1996). Results show that there is no guarantee from any of these algorithms because they do not consider space. More experimental evaluations will be presented in the next subsection.

It must be emphasized that the benchmark used for evaluating the results is the original line (but not the line at a smaller scale). Therefore, these algorithms have nothing to do with scale, so there is no guarantee of their results for multi-scale representation.

5.5.2 PERFORMANCE OF POINT-REDUCTION ALGORITHMS

There are some problems with most point-reduction algorithms (Visvalingham and Whyatt, 1990; Saalfeld, 1999). The following are the most serious ones:

1. The final result will be dependent on the initial set of critical points.
2. There is no guarantee of self-intersections or cross-intersections between lines.
3. Small spikes are always retained so that large shape distortion may be created.
4. The points that are retained are not necessarily critical to the line representation.

The first point is first discussed in Section 5.3.2 and illustrated by Figure 5.11. The self-intersection problem is illustrated in Figure 5.18. The cross-intersections between lines are illustrated in Figure 5.19. Large shape distortion may also be caused by such algorithms because they tend to retain the spiky points, as illustrated in Figure 5.20.

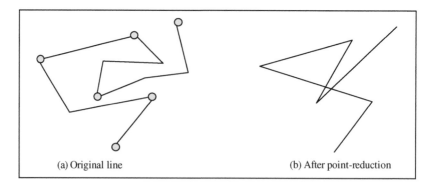

(a) Original line   (b) After point-reduction

FIGURE 5.18 Self-intersections created by point-reduction algorithms.

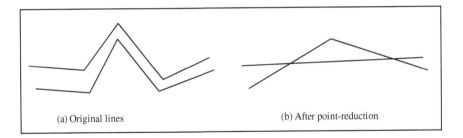

(a) Original lines   (b) After point-reduction

FIGURE 5.19 Cross-intersections between lines created by point-reduction algorithms.

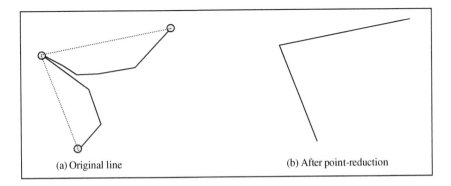

(a) Original line   (b) After point-reduction

FIGURE 5.20 Large distortion created by point-reduction algorithms.

All these point to the conclusion that these algorithms are not suitable for transformations in scale dimension, as pointed out by Li (1993) and supported by Christensen (1999, 2000).

5.6 ATTEMPTS TO IMPROVE POINT-REDUCTION ALGORITHMS

Point reduction through curve approximation and critical point detection is a topic of great interest to many researchers. The literaturehas been flooded in recent years with articles on the subject (e.g., Cornic, 1997; Neumann and Teisseron, 2002). It is impossible to include all of them. For example, algorithms working in the frequency domain (e.g., Persoon and Fu, 1977; Lee et al., 1993; Antoine et al., 1997) are all omitted here. This is certainly a shortcoming. However, as emphasized in many places, point reduction is not a main issue in multi-scale spatial representation anymore, so it is not too unwise to omit such materials from this text.

However, it may be worth including some material about the improvement of the performance of point-reduction algorithms, as there are some problems associated with such algorithms.

5.6.1 Attempts to Avoid Topological Conflicts

To solve intersection problem, two options are possible. One solution is to develop some sort of postprocessing algorithm to remove the conflicts, and the alternative is to include some kind of conditions for such topological conflicts.

Muller (1990) has developed a geometric procedure for the removal of two types of spatial conflicts: point-to-point and point-to-line collisions. de Berg et al. (1995, 1998) have developed an algorithm to avoid spatial conflicts by considering the chain (any sequence of vertices and edges starting and ending at a junction) as a graph and introducing a number of constraints. Saalfeld (1999) has made use of the dynamically updated convex hull data structure to efficiently detect and remove potential topological conflicts (i.e., self-intersection and cross-intersection), as the displacements caused by point-removal always fall within the convex hull of the original line.

Zhang and Tian (1997) have tried to modify the Douglas–Peucker algorithm to ensure that the new line segments are moved to one side of original line, for use in marine charts.

5.6.2 Attempts to Make Algorithms Robust

To make algorithms robust, many researchers have tried to integrate corner detection with polygonal approximation. These hybrid algorithms might also be classified into two groups:

1. Corner detection followed by a split-and-merge process.
2. Corner detection followed by a progressive splitting process.

Ansari and Delp (1991) make use of the points with extreme curvature as the initial set of critical points and then apply a split-and-merge process. In their algorithm a Gaussian smoothing filter is first applied to the original curve; then a set of points with maximum positive and negative curvature on the Gaussian-smoothed lines is detected, and finally a split-and-merge process is followed to detect more critical points. There is a trade-off in the selection of the width for the Gaussian filter. Too large a width will remove all small variations, and too small a width will create false concavities and convexities.

Wu and Wang (1993) also make use of the points with extreme curvature as the initial set of critical points. Instead of a split-and-merge process, they apply the progressive splitting process. That is, each curve segment between two consecutive critical points is further split until certain criteria are met. Both the initial set of critical points and those detected during progressive splitting are considered critical points.

There are many possible combinations, because there are so many corner detection and polygon approximation algorithms. Wu and Wang (1993) found that a combination of the Rosenfeld–Johnston angle detection algorithm and the Ramer polygonal approximation algorithm produces the most satisfactory results.

Another development to make the algorithms more robust is the employment of a smoothing process before corner detection or polygonal approximation. Gaussian smoothing is particular popular (e.g., Thapa, 1988b; Rattarangsi and Chin, 1992;

Pei and Lin, 1992). Pei and Lin (1992) have claimed that this technique is capable of detecting stable cardinal points accurately and that the resultant critical points will not change under scaling, translation, and rotation. However, the positions of the critical points are not on the original line anymore, as the algorithms work on the smoothed line.

5.6.3 ATTEMPTS TO MAKE ALGORITHMS SELF-ADAPTIVE

Another type of improvement is to make the algorithm self-adaptive. In such algorithms, a line is first split into pieces as preprocessing and then point-reduction is applied, but with different criteria for different pieces, leading to the concept of "piece-wise linear approximation" (e.g., Aoyama and Kawagoe, 1991; Sato, 1992). Wang and Muller (1998) present a piecewise algorithm for approximation of a road. The algorithm split a road into a few bands based on an analysis of shape characteristics. Wang and Muller (1993) have also presented an algorithm specifically designed for coastlines by taking into consideration the semantics of rivers.

Ansari and Huang (1991) applied an adaptive Gaussian filter with a width proportional to the region-of-support to smooth the curve. The region-of-support, as suggested by Teh and Chin (1989), also needs to be determined for each point (see Section 5.4.2). As a result, this algorithm is less sensitive to noise, but its complexity has increased dramatically.

5.6.4 ATTEMPTS TO MAKE ALGORITHMS MORE EFFICIENT

Efforts have also been made to improve the efficiency (i.e., the speed) of point-reduction algorithms. Hershberger and Snoeyink (1992) noticed that the most time-consuming process is to find the point that deviates most from the line segment under consideration. They made use of the convex hull to speed up the search process because the most deviated point must be inside the convex hull.

Vaughan et al. (1991) have tried to develop parallel algorithms to speed up the process.

REFERENCES

Ansari, N. and Delp, E., On detecting dominant points, *Pattern Recognition*, 24(5), 441–451, 1991.

Ansari, N. and Huang, K.-W., Non-parametric dominant point detection, *Pattern Recognition*, 24(9), 849–862, 1991.

Antoine, J.-P., Barache, D., Cesar, R. M., and da Fontoura, L., Shape characterization with the wavelet transform, *Signal Processing*, 62, 265–290, 1997.

Aoyama, H. and Kawagoe, M., A piecewise linear approximation method preserving visual feature points of original figures, *CVGIP: Graphic Models and Image Processing*, 53(5), 435–446, 1991.

Attneave, F., Some informational aspects of visual perception, *Psychological Review*, 61(3), 183–193, 1954.

Christiansen, A. H., Cartographic line generalization with waterlines and medial-axes. *Cartography and Geographic Information Science*, 26(1): 19–32, 1999.

Christiansen, A. H., Line generalization by waterline and medial-axis transformation: success and issues in an implementation of Perkel's proposal. *The Cartographic Journal*, 26(1): 19–32, 2000.

Cornic, P., Another look at the dominant point detection of digital curves, *Pattern Recognition Letters*, 18, 13–25, 1997.

de Berg, M., van Kreveld, M. and Schirra, S., A new approach to subdivision simplification, *Auto-Carto*, 12, 19–88, 1995.

de Berg, M., van Kreveld, M., and Schirra, S., Topologically correct subdivision simplification using the bandwidth criterion, *Cartography and Geographic Information Science*, 25, 243–257, 1998.

Douglas, D. H. and Peucker, T. K., Algorithms for the reduction of the number of points required to represent a digitized line or its caricature, *Canadian Cartographer*, 10(2), 112–122, 1973.

Duda, R. O. and Hart, P. E., *Pattern Classification and Scene Analysis*. Wiley, New York, 1973.

Freeman, H., Shape description via the use of critical points, *Pattern Recognition*, 10, 159–166, 1978.

Freeman, H. and Davis, L. S., A corner-finding algorithm for chain-coded curves, *IEEE Transactions on Computers*, 26, 297–303, 1977.

Hershberger, J. and Snoeyink, J., Speeding up the Douglas-Peucker line-simplification algorithm, in *Proceedings of the 5th International Symposium on Data Handling*, 1992, pp. 134–143.

Lang, T., Rules for robot draughtsmen, *Geographic Magazine*, 42(1), 50–51, 1969.

Lee, J.-S., Sun, Y.-N., Chen, C.-H., and Tsai, C.-T., Wavelet based corner detection, *Pattern Recognition*, 26(6), 853–865, 1993.

Li, Z. L., An algorithm for compressing digital contour data, *Cartographic Journal*, 25(2), 143–146, 1988.

Li, Z. L., Some observations on the issue of line generalisation, *Cartographic Journal*, 30(1), 68–71, 1993.

Li, Z. L., An examination of algorithms for detection of critical points on digital lines, *Cartographic Journal*, 32(2), 121–125, 1995.

Li, Z. L. and Openshaw, S., A comparative study of the performance of manual generalization and automated generalizations of line features. *AGI 1992–1993* Year Book. The Association for Geographic Information, London, 502–513, 1992.

McMaster, R. B., A statistical analysis of mathematical measures for line simplification, *American Cartographer*, 13, 103–116, 1986.

McMaster, R. B., Automated line generalisation, *Cartographica*, 24(2), 74–111, 1987.

Muller, J.-C., The removal of spatial conflicts in the line generalisation, *Cartography and Geographic Information Systems*, 17(2), 141–149, 1990.

Nako, B. and Mitropoulos, V., Local length ratio as a measure of critical point detection for line simplification, *5th ICA Workshop on Progress in Automated Map Generalisation*, April 28–30 2003, Paris (http://www.geo.unizh.ch/ICA/).

Neumann, R. and Teisseron, G., Extraction of dominant points by estimation of contour fluctuations, *Pattern Recognition*, 35, 1447–1462, 2002.

Opheim, H., Smoothing a digitized curve by data reduction methods, *Eurographics'81*, 127–135, 1981.

Pavlidis, T., A review of algorithms for shape analysis, *Computer Graphics and Image Processing*, 7, 243–258, 1978.

Pavlidis, T. and Horowitz, S. L., Segmentation of plan curves, *IEEE Transaction on Computers*, 23, 860–870, 1974.

Pei, S.-C. and Lin, C.-N., The detection of dominant points on digital curves by scale-space filtering, *Pattern Recognition*, 25(11), 1307–1314, 1992.

Persoon, E. and Fu, K.-S., Shape discrimination using Fourier descriptors, *IEEE Transactions on Systems, Man and Cybernetics*, 7(3), 170–179, 1977.

Ramer, U., An iterative procedure for polygonal approximation of plane curves, *Computer Graphics and Image Processing*, 1, 244–256, 1972.

Rattarangsi, A. and Chin, R. T., Scale-based detection of corners of planar curves, *IEEE Transactions on Pattern Analysis and Machine Intelligence,* 14(4), 430–449, 1992.

Ray, B. K. and Ray, K. S., A new approach to polygonal approximation, *Pattern Recognition Letters*, 12(4), 229–234, 1991.

Reumann, K. and Witkam, A. P. M., Optimizing curve segmentation in computer graphics, in *Proceedings of International Computing Symposium*, Amsterdam, North-Holland Publishing Company, 1974, pp. 467–472.

Rosenfeld, A. and Johnston, E., Angle detection of digital curves, *IEEE Transactions on Computers*, 22, 875–878, 1973.

Rosenfeld, A. and Weszka, J. S., An improved method of angle detection on digital curves, *IEEE Transactions on Computers*, 24, 940–941, 1975.

Saalfeld, A., Topologically consistent line simplification with the Douglas-Peucker algorithm, *Cartography and Geographic Information Science*, 26(1), 7–18, 1999.

Sankar, P. V. and Sharma, C. V., A parallel procedure for detection of dominant points on a digital curve, *Computer Graphics and Image Processing*, 7, 403–412, 1978.

Sato, Y., Piecewise linear approximation of plane curves by perimeter optimization, *Pattern Recognition*, 25(12), 1535–1543, 1992.

Sklansky, J. and Gonzalez, V., Fast polygonal approximation of digitized curves, *Pattern Recognition*, 12, 327–331, 1980.

Teh, C.-H. and Chin, R. T., On the detection of dominant points on digital curves, *IEEE Transactions on Pattern Analysis and Machine Intelligence*, 11(8), 859–872, 1989.

Thapa, K., A review of critical points detection and line generalization algorithms, *Surveying and Mapping*, 48(3), 185–205, 1988a.

Thapa, K., Critical points detection and automated line generalization in raster data using zero-crossings, *Cartographic Journal*, 25(1), 58–68, 1988b.

Vaughan, J., Whyatt, D., and Brookes, G., A parallel implementation of the Douglas–Peucker line simplification algorithm, *Software: Practice and Experience*, 21(3), 331–336, 1991.

Visvalingham, M. and Whyatt, J., The Douglas-Peucker algorithm for line simplification: re-evaluation through visualization, *Computer Graphics Forum*, 9, 213–228, 1990.

Visvalingham, M. and Whyatt, J., Line generalization by repeated elimination of points, *Cartographic Journal*, 30(1), 46–51, 1993.

Wall, K. and Danielsson, P. E., A fast sequential method for polygon approximation of digitized curves, *Computer Vision, Graphics and Image Processing*, 28, 220–227, 1984.

Wang, Z. and Muller, J. C., Complex coastline generalization, *Cartography and Geographic Information Systems,* 20(2), 96–106, 1993.

Wang, Z. and Muller, J.C., Line generalization based on analysis of shape characteristics, *Cartography and Geographic Information Systems,* 25(1), 3–15, 1998.

White, E., Assessment of line-generalization algorithms using characteristics points, *American Cartographer*, 12(1), 17–27, 1985.

Wu, W.-Y. and Wang, M.-J., Detecting the dominant points by curvature-based polygonal approximation, *CVGIP: Graphical Models and Image Processing*, 55(2), 79–88, 1993.

Zhang, L. and Tian, Z., Refinement of Douglas-Peucker algorithm to move the segments toward only one side, in *Proceedings of 18th International Cartographic Conference*, Stockholm, Sweden, 1997, pp. 830–835.

6 Algorithms for Smoothing of Individual Line Features

In the previous chapter, a selection of algorithms for line point-reduction was discussed. In this chapter, some algorithms for *line smoothing* will be discussed.

6.1 SMOOTHING OF A LINE: AN OVERVIEW

Smoothing is a *filtering process* to reduce small variations, and it is not necessarily directly related to scale change. Smoothing can be performed in the space domain or in the frequency domain.

In the frequency domain the basic idea is to cut off the high frequency components of a given line. *Fourier transform* and *wavelet transform* are two commonly used tools. More recently, a data-driven decomposition method, called *empirical mode decomposition (EMD)*, has been developed (Huang et al., 1998) and used for smoothing data to obtain trends (Li et al., 2004). In the space domain two approaches are widely used: *moving averaging* and *curve fitting*. Three types of curve fitting can be distinguished: *least-squares fitting, exact fitting,* and *energy-balancing fitting.* Therefore, line smoothing algorithms and operators can be categorized as shown in Figure 6.1:

6.2 SMOOTHING BY MOVING AVERAGING IN THE SPACE DOMAIN

Moving averaging is widely used for interpolation and smoothing. In spatial representation, Tobler (1966) applied moving averaging for smoothing a set of three-dimensional (3-D) data in the early 1960s. The basic idea is to apply an averaging filter to each point sequentially. The averaging can be either weighted or not weighted.

6.2.1 SMOOTHING BY SIMPLE MOVING AVERAGING

Let a line, consisting of N points, be expressed as $\{(x_1, y_1), (x_2, y_2), \ldots (x_N, y_N)\}$. The basic formula for the simple moving averaging is then

$$y_i' = \frac{\sum\limits_{j=-k}^{k} y_{i+j}}{2k+1} \tag{6.1}$$

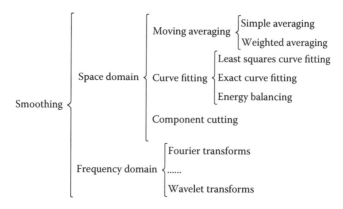

FIGURE 6.1 A classification of line smoothing algorithms.

where y_i is the new y coordinate for the ith point in the data set, k is the half size of the window, and j is the numbering of the point in the window (to the center point of the window $[-k, k]$). This means that to obtain the new y value (y_i') for the ith point in the data set, all the points within the window $[i-k, i+k]$ are used to compute the average.

The algorithm for simple averaging works as follows:

1. Determine the window size $[-k, k]$.
2. Start the process from $i = k + 1$.
3. For a data point i, compute the new y value (y_i') for each point, according to Equation 6.1.
4. Move to the next point and repeat step 3 until $i = N - k$.

Figure 6.2 illustrates the result of simple moving averaging for the following data set:

$$\{(1,9), (2,5), (3,3), (4,6), (5,14), (6,12), (7,9), (8,11)\}$$

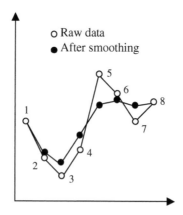

FIGURE 6.2 An example of moving averaging.

In this example there are eight points, that is, $N = 8$. A window of three points is selected, that is, $k = 1$. The two end points remain unchanged. The smoothing process starts from point 2. For $i = 2$, the y coordinates of points 1, 2, and 3 are used to make an average, and the result y_2' is $(9 + 5 + 3)/3 = 17/3$. Then one moves to the next point, $i = 3$. In this case, the y coordinates of points 2, 3, and 4 are used to make an average, that is, $y_3' = (5 + 3 + 6)/3 = 14/3$. The process continues until point 7 is smoothed.

6.2.2 SMOOTHING BY WEIGHTED MOVING AVERAGING

Different weights may be assigned to different points within the window $[i - k, i + k]$, leading to the term *weighted moving averaging*. If each point is given an equal weight, it is called simple averaging. However, it is a common practice to give a greater weight to a point closer to the central point, that is, point i. To understand the idea of variable weighting, different kinds of weighting functions can be employed. The simplest is the distance (d) function as follows:

$$W_j = \frac{1}{d_j^r} \tag{6.2}$$

where d_j is the distance of the jth point to the center of the window, and r is the power of the distance. If $r = 1$, W is inversely proportional to d.

Another popular weighting function is the *Gaussian function*, as follows:

$$G(x) = \frac{1}{\sqrt{2\pi}\sigma} e^{-\frac{x^2}{2\sigma^2}} \tag{6.3}$$

where σ is the standard deviation of the distribution that is assumed to have a zero mean (μ), that is, with a center line $x = 0$. Figure 6.3 is such a normal distribution with $\sigma = 1$. The probability values to different locations of the function (such as -3σ, -2σ, $-\sigma$, 0, σ, 2σ, 3σ,) can be used as the weights. For example, if the size of a window is 7, a possible kernel from this function is (0.006, 0.061, 0.242, 0.383, 0.242, 0.061, 0.006). However, this may require data points with equal spacing.

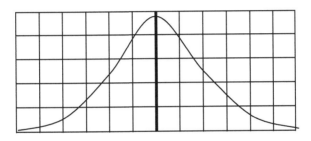

FIGURE 6.3 A normal distribution with $\mu = 0$, $\sigma = 1$.

The mathematical equation for weighted averaging is as follows:

$$y_i' = \frac{\sum\limits_{j=-k}^{k} W_j\, y_{i+j}}{\sum\limits_{j=-k}^{k} W_j} \tag{6.4}$$

where y_i' is the new y coordinate for the ith point in the data set, k is the half size of the window, j is the numbering of the point in the window (to the center point of the window), and W_j is the weight for the jth point in the window $[-k, k]$.

The algorithm for weighted averaging works as follows:

1. Determine a weighting function.
2. Determine the window size $[-k, k]$.
3. Start the process from $i = k + 1$.
4. For each point in the data (point i), compute a set of weights, one for each of the points in the window (i.e., W_j for point $[i+ j]$).
5. Compute the new y value y_i' for each point according to Equation 6.4.
6. Move to the next point and repeat steps 4 and 5 until $i = N - k$.

6.3 SMOOTHING BY CURVE FITTING IN THE SPACE DOMAIN

For a given set of data, there will be many kinds of fitting techniques. This section introduces three of them.

6.3.1 SMOOTHING BY BEST FITTING: LEAST-SQUARES

Curve fitting is best fitting a curve function into a set of data points $\{(x_1, y_1), (x_1, y_1), \dots (x_N, y_N)\}$. The commonly used function is a *polynomial* consisting of one or more terms as follows:

$$y = a_0 x^0 + a_1 x^1 + a_2 x^2 + \cdots + a_n x^n \tag{6.5}$$

The names of the polynomials at different orders (or degrees) are shown in Table 6.1. For example, the second order polynomial function $y = a_0 x^0 + a_1 x^1 + a_2 x^2$ is called the quadratic function.

TABLE 6.1
Names of Polynomials at Different Degrees

Order	0	1	2	3	4
Name	Plane	Linear	Quadratic	Cubic	Quartic

This is the linear family of curves. There are also exponential and power law families, as shown in Equation 6.6 and Equation 6.6. However, the polynomials are still the most popular because of linearity.

$$\text{Exponential:} \quad y = a\,e^{bx} \tag{6.6}$$

$$\text{Power:} \quad y = a\,x^b \tag{6.7}$$

If the number of known points is more than the number of coefficients in the selected polynomial function, then there is an infinite number of fittings, and one must select a best fitting or optimum solution. The commonly used condition for such an optimization is the *least-squares* solution. That is, the sum of the squares of the residuals at the known points is to be a minimum:

$$\sum_{i=0}^{N} r_i^2 = min \tag{6.8}$$

where r_i is the residual at point i, and N is the total number of points used in the fitting. The algorithm for *least squares curve fitting* works as follows:

1. Divide a line into pieces if it is too long.
2. Select a mathematical function for curve fitting, for example, M terms from Equation 6.5.
3. Apply least-squares regression to the set of data (with N points) to obtain the set of coefficients $\{a_0, a_1, \ldots, a_{M-1}\}$ $\{a_0, a_1, \ldots, a_{M-1}\}$ of the curve function.
4. Use the coefficients that have just been obtained in step 3 to compute the new y coordinates of each point.

Figure 6.4 illustrates the least-squares fitting with a cubic function:

$$y = a_0 x^0 + a_1 x^1 + a_2 x^2 + a_3 x^3 \tag{6.9}$$

The original data set consists of eight points: $\{(1,9), (3,6), (4,3), (7,6), (9.14), (10,12), (13,9), (15,11)\}$. The four coefficients obtained by least-squares regression are $(12.0815, -4.1177, 0.6945, -0.0288)$. Thus, the resultant smoothing function for curve fitting is

$$y = 12.0815 - 4.1177x + 0.6945x^2 - 0.0288x^3$$

which is represented by a curved line in Figure 6.4. The residuals in y coordinates at these eight points are $(0.3705, 0.7984, -1.8801, -1.4113, 3.7160, 0.4426, -3.6526, 1.6165)$.

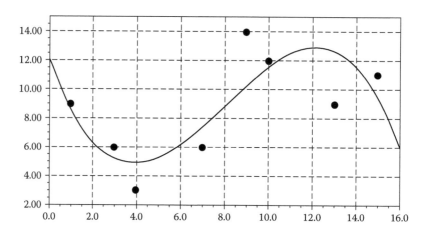

FIGURE 6.4 Cubic curve fitting by least-squares solution.

6.3.2 SMOOTHING BY EXACT FITTING: CUBIC SPLINE

There are different types of *spline,* such as Bezier curves, B-spline, tension spline, and cubic spline. The first two are similar to the least squares fitting in that they do not require the curve to pass through the given set of points. However, cubic and tension splines require the curve to pass through all the given points one by one sequentially and smoothly, leading to the term *exact fitting.* In this section, only the cubic spline will be discussed.

In *cubic spline,* a cubic curve is formed between each pair of control points. From Equation 6.9, it can be seen that there are four coefficients in the cubic function and there is a need of four equations to solve these four coefficients. With two control points, one can have only two equations formed from the cubic function. Therefore, there is a need of another two equations.

Let the coordinates of the two nodes $(i - 1)$ and i be (x_{i-1}, y_{i-1}) and (x_i, y_i) (see Figure 6.5); the cubic function is $f(x)$. Then the two equations for the two control points are

$$y_i = f(x_i)$$
$$y_{i-1} = f(x_{i-1})$$

(6.10)

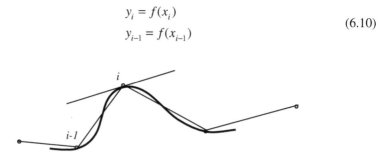

FIGURE 6.5 Fitting a cubic function between two points $(i - 1)$ and i.

The other two conditions could be that the first derivatives at $(i - 1)$ and i should be equal to the given values:

$$y_i' = f'(x_i)$$
$$y_{i-1}' = f'(x_{i-1})$$

(6.11)

With these four equations in Equations 6.10 and 6.11, the four coefficients of Equation 6.9 for the piece between points $(i - 1)$ and i, that is, $\{a_{i,0}, a_{i,1}, a_{i,3}, a_{i,4}\}$ can then be obtained.

The four coefficients obtained from these four equations are only for the piece of curve between points $(i - 1)$ and i. For another neighboring piece between points i and $(i + 1)$, another set of coefficients, $\{a_{i+1,0}, a_{i+1,1}, a_{i+1,3}, a_{i+1,4}\}$, will be obtained. To make sure that the curve pieces join together smoothly, the first and second derivatives at the end of one piece must be equal to those at the beginning of the next piece. These requirements usually serve as the other two equations. The second derivatives at the end points of each piece are commonly set to zero. As a result, computation of cubic spline coefficients essentially means the resolution of simultaneous equations. For readers who are interested in mathematical deduction, the text provided on Weisstein's (2005) text on the Internet site serves as an excellent introduction.

The algorithm for *cubic curve fitting* works as follows:

1. Select the cubic function for curve fitting (Equation 6.9).
2. Use data points as controls and derivatives as constraints to build a system of simultaneous equations.
3. Solve the equations and obtain a set of four coefficients for each piece (between two control points), for example, $\{a_{i,0}, a_{i,1}, a_{i,3}, a_{i,4}\}$, for the piece between points $(i + 1)$ and i.

Figure 6.6 shows the cubic spline of the same set of data as used for Figure 6.4. It can be seen that the curve can wriggle in quite unexpected ways.

6.3.3 SMOOTHING BY ENERGY MINIMIZATION: SNAKES

In a least-squares solution, the sum of squares of the residuals at control points is minimized. In signal processing, another solution is to minimize the so-called energy. This technique is called *snakes*.

Snakes was first introduced by Kass et al. in 1987. A model called *tangent angle function snakes* (TAFUS) was developed by Borkowski et al. (1999) specifically for cartographic purposes. Burghardt and Meier (1997) introduced the snakes technique for cartographic displacement. Steiniger and Meier (2004) have compared the conventional snakes model and TAFUS and recommended the traditional model.

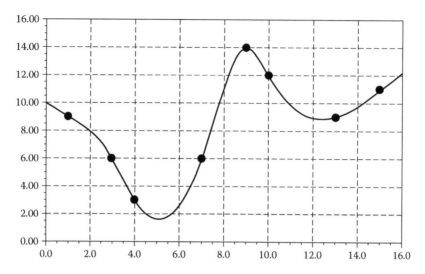

FIGURE 6.6 Cubic curve fitting by spline.

A snake model is thought to balance the internal energy and external energy of a line. The former is meant to smooth the line, while the latter is meant to react against the smoothing. The mathematical function is as follows (Steiniger and Meier, 2004):

$$E_{Snakes} = \int_0^1 (E_{Ext} + E_{Int}) dl = min \qquad (6.12)$$

where E_{Snakes} is the energy of the snakes, E_{Ext} is the external energy describing the external influence or the interaction with other features, E_{Int} is the internal energy describing the shape and characteristics of the line, and l is the arc length. The energy function defined over the curved line, E_{Snakes}, is minimized by changing the curve's shape and position, which makes the curve move across the space like a snake. In the case of line smoothing, external energy may be set to a constant (Steiniger and Meier, 2004).

The concept of energy was introduced by Young and his collaborators (Young et al., 1974) to describe line complexity as a global measure of complexity. The bending energy, or boundary energy, is defined as follows:

$$E_{Bending} = \frac{1}{P} \int k(l)^2 ds \qquad (6.13)$$

where $E_{Bending}$ is the bending energy or boundary energy, P is the contour perimeter, and $k(l)$ is the curvature at position l (the arc length).

This measure was inspired by the concept of elasticity in physics and is used to describe the amount of energy required to transform a given close contour to a circle with the same perimeter. The internal energy of the snakes model is an analogy of the bending energy by Young et al. (1974), as

$$E_{Int} = \frac{\alpha}{2}\left|\frac{dg}{dl}\right|^2 + \frac{\beta}{2}\left|\frac{d^2g}{dl^2}\right|^2 = \frac{1}{2}\left(\alpha|g'|^2 + \beta|g''|^2\right) \tag{6.14}$$

where α is a parameter controlling the degree of elasticity, and β is a parameter controlling the degree of flexibility.

The important issue is to select optimum values for α and β. A large value for α will increase the internal energy of the snake as it stretches more and more, while a large value for β will increase the internal energy of the snake as it develops more curves. Figure 6.7 shows the effects of α and β values on the final results. It can be seen that a β much larger than α will produce a strongly smoothed result. Theoretically speaking, one could set appropriate values for both α and β to obtain a smoothed line in one step by variational calculus (Kass et al., 1987). However, it is

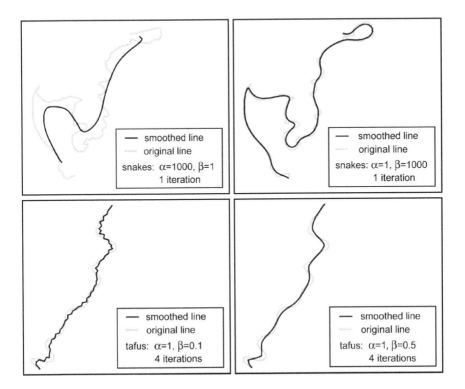

FIGURE 6.7 Effects of α and β values on the result of line smoothing (Reprinted from Steiniger and Meier, 2004. With permission.).

not easy and indeed requires a lot of experience. It is advised to make the process iterative, for example, using the Greedy algorithm by Williams and Shah (1990). Steiniger and Meier (2004) recommend:

1. The initial values are set as $\alpha = \beta = 1$.
2. Make the β varied.
3. Stop when the degree of smoothing is achieved by using curvature or number of iterations as the criterion.

It should be noted here that the data point of the original line should be represented in a parametric form, that is,

$$g(l) = \begin{cases} X = f_x(l) \\ Y = f_y(l) \end{cases} \tag{6.15}$$

In some models such as the TAFUS model, equal arc length is required. Figure 6.8 shows the parametric representation of a set of line points with equal arc length.

The algorithm for line smoothing using the snakes model works as follows:

1. Divide a line into pieces if it is too long.
2. Separate the coordinates of the line into two sets, one set for X and the other for Y.
3. Represent the X and Y coordinates separately in parametric form, that is $X = f_x(l)$ and $Y = f_y(l)$.
4. Set a constant for external energy.
5. Set initial values for internal energy, such as $\alpha = \beta = 1$.
6. Make the β varied to obtain a smoothed result.
7. Repeat step 5 until a criterion is met, which could be a value of curvature or number of iterations.

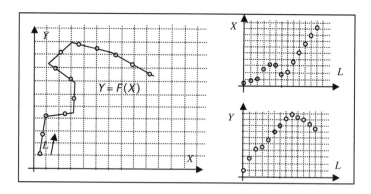

FIGURE 6.8 Representation of a polyline in parametric form with equal arc length.

6.4 SMOOTHING BY FREQUENCY CUTTING IN THE FREQUENCY DOMAIN

6.4.1 SMOOTHING BY FOURIER TRANSFORMS

The idea of the *Fourier transform* is to represent data as a superposition of sine and cosine functions. The Fourier transform has been widely used as a filtering technique in many disciplines, especially with the introduction of *fast Fourier transform* (FFT) (Cooley and Tukey, 1965) and the consequent subroutine development (Brenner, 1969). Such subroutines have been implemented in almost all software packages related to computations.

The principle of applying FFT for line smoothing is similar to the filtering process in other disciples. The procedure works as follows:

1. Separate the coordinates of the line into two sets, one set for X and the other for Y.
2. Resample each set into equidistant points.
3. Apply an FFT to each set to obtain a representation in frequency.
4. Cut off undesirable frequencies.
5. Apply inverse FFT to obtain a filtered set, one for X and one for Y.
6. Reconstruct the line from the filtered X and Y sets.

In practice, this procedure does not work well. Boutoura (1989) has made use of the slope values instead of the actual X and Y coordinates, because slope is considered to be a quantity intrinsically related to the geometry of a line, as it describes the shape behavior of the line with respect to its curvature changes. The procedure is as follows:

1. Resample the original line $y = f(x)$ into equidistant points.
2. Map the original line into its slope counterpart, that is, compute a series of slope values as a discrete function of arc length, $g(l)$.
3. Apply FFT to the slope function $g(l)$ to obtain frequency spectrum $G(u)$.
4. Cut off undesirable frequencies from $G(u)$ to obtain a new spectrum $G'(u)$.
5. Apply inverse FFT to the new spectrum $G'(u)$ to obtain the smoothed slope function $g'(l)$.
6. Reconstruct a line $y' = f'(x')$ from the smoothed slope function $g'(l)$.
7. Apply a best fitting $y' = f'(x')$ into $y = f(x)$ to obtain a smoothed line $y' = f'(x)$. This is because there will be more than one $y' = f'(x')$ corresponding to $g'(l)$. The constraint used for best fitting is $[(x_i' - x_i)^2 + (y_i' - y_i)^2 = \min$.

The mathematical expressions for this procedure are listed in Table 6.2. The degree of filtering is dependent on the value of N_C (cut-off frequency). If N_C is closer to N_{Nyq} (Nyquist frequency), then the degree of smoothing is low. If N_C deviates more from N_{Nyq}, then the line will be smoother.

TABLE 6.2
Mathematical Expressions for Fourier-Based Line Smoothing

Feature or Process	Mathematical Equation
Function of a line	$y = f(x)$
Slope of point j	$Slope_j = \left(\dfrac{dy}{dx}\right)_j$, $j = 1, 2, \ldots, M\text{-}1$
Fourier transform	$R_k = \dfrac{\sum\limits_{j=0}^{N-1} x_j \cos(2njk/N)}{N}$, $I_k = \dfrac{\sum\limits_{j=0}^{N-1} x_j \sin(2njk/N)}{N}$ $j = 0, 1, \ldots, N\text{-}1$ [Where, N is the number of real values (slopes)]
Inverse FT	$x_j = \sum\limits_{k=0}^{N-1} \left[R_k \cos(2njk/N) - I_k \sin(2njk/N) \right]$
Filter	$F_k = \begin{cases} 1-(k/N_C)^2, & k=1,2,\ldots,N_C-1 \\ 0 & k = N_C, N_C+1,\ldots, \end{cases}$ where N_C is the cut-off frequency and $0 \le k \le Nyq$ (Nyquist frequency, equal to $[N-1]/2$ if N is odd and equal to $[N/2]$ if N is even).
Filtering process	$R'_k = F_k R_k$ $I'_k = F_k I_k$

Boutoura (1989) suggests relating the N_C value to the source map scale M_S and target map scale M_T. Suppose the source map is digitized with a discretization step Δ_S; the corresponding value on the target map is

$$\Delta_T = \Delta_S \times (M_T/M_S) \tag{6.16}$$

The wavelength corresponding to the cut-off frequency N_C is

$$\Delta\lambda_k = \Delta_T \tag{6.17}$$

where $\Delta\lambda_k$ is the difference between the wavelength of kth wave and that of $(k-1)$th wave in the Fourier transform.

Figure 6.9 is an example used by Boutoura (1989), showing the filtering of a closed contour line from 1:5,000 to 1:10,000. It seems that too much has been filtered out for a two times reduction in scale. More importantly, no intensive test has been carried out against the manually generalized result, and it is still unknown whether or not this method is capable of producing results, which is compatible to the existing manual work. In other words, it is still unknown whether or not this method is applicable in practice.

6.4.2 Smoothing by Wavelet Transforms*

The fundamental idea behind *wavelet transforms* is to analyze data according to scale. Wavelets is basically a windowing technique with variably sized regions that breaks

---

* This section is largely extracted from Balboa and Lopez, 2000. With permission.

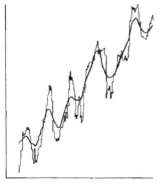

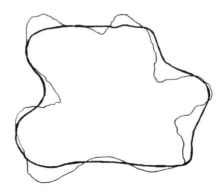

(a) Filtering the slope function   (b) Superimposition of smoothed line onto the original

FIGURE 6.9 Filtering of a contour by Fourier transform (Reprinted from Boutoura, 1989. With permission.).

up a signal into shifted and scaled versions of a mother wavelet. Hence, the wavelet spectrum is a plot of wavelet coefficients (amplitude) over time and scale axes. In other words, it yields the energy as a function of time and scale (frequency). Although the first mention of wavelets appeared in an appendix to the thesis by A. Haar in 1909, the method was not widely used until 1985, when Stephane Mallat gave wavelets an additional jump-start through his work in digital signal processing (Graps, 1995).

The *Haar wavelet* technique takes averages and differences of a signal, for example, a set of line points. That is, in each step, two sets of values are produced, one for averages and the other for differences. The averages are the smoothed versions of the data and the differences are called *wavelet coefficients*. The averages are the input for the next iteration of the transform. This process continues until only one average and one coefficient are computed. Therefore, in the end, the number of coefficients is an increasing power of 2 (i.e., 2^0, 2^1, 2^2, 2^3, 2^4, etc.), as shown in Figure 6.10.

To make the principle of Haar wavelet transform clear, the following set of data with eight samples is used:

$$f = (2, 4, 2, 0, 4, 10, 7, 3)$$

which is the example by Balboa and Lopez (2000). There are four pairs in the data set. From Figure 6.10 it can be seen that three rounds of computation are required to complete the transform. In the first round, four averages and four coefficients are

Last data average	Last (2^0) Coefficient (Difference)	2 (2^1) Coefficients (Differences)	4 (2^2) Coefficients (Differences)	8 (2^3) Coefficients (Differences)	16 (2^4) Coefficients (Differences)	...

FIGURE 6.10 The form of wavelet coefficients.

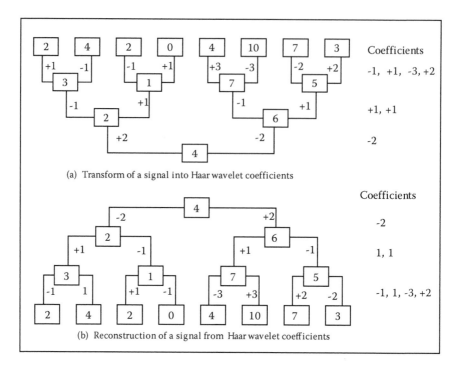

(a) Transform of a signal into Haar wavelet coefficients

(b) Reconstruction of a signal from Haar wavelet coefficients

FIGURE 6.11 Decomposition and reconstruction by Haar basis (Modified from Balboa and Lopez, 2000. With permission.).

obtained. These four averages are the inputs for the second iteration. From these four, two averages and two coefficients are obtained. In the third round, only one average and one coefficient are obtained. The whole process is illustrated in Figure 6.11. The final result is one average plus a set of 7 ($= 2^0 + 2^1 + 2^2$) coefficients with, that is, [4, −2,1,1, −1,1, −3,2. Figure 6.11b shows the reconstruction from this set of coefficients.

Wavelet has also been used for filtering of line features (e.g., Plazanet et al., 1995; Balboa and Lopez, 2000). The procedure described by Balboa and Lopez (2000) is as follows:

1. Curved line $y = f(x)$ is transformed into a frequency function $g(l,c)$. Here l is the arc length and c is the curvature evolution, along the line.
2. Resample the $g(l,c)$ into 2^j initial values with equal length intervals. The coordinates of the points need to be resampled in the same way.
3. Decompose the resampled frequency representation into wavelet functions (e.g., Haar's wavelet function) $\phi(c)$.
4. Erase the coefficients of the wavelets with small values (smaller than a threshold), so as to obtain a new set of coefficients $\phi'(c)$.
5. Transform the filtered result $\phi'(c)$ back to the space domain by inverse transform to obtain a new representation by length and curvature $g'(l,c)$.
6. Construct the line $y = f'(x)$ from $g'(l,c)$ in the end.

One basic concept in wavelet is the orthogonal basis of the vector space, upon which any other vector of this space can be expressed by a linear combination of this set of basic vectors. This set of basic vectors can be compared with the vector (1,0) and (0,1) in 2-D Euclidean space, in the sense that every 2-D vector (x,y) is a combination of the vector (1,0) and (0,1). Suppose there are 2^j curvature values to be generalized; one needs to have an orthogonal space with j dimension V^j. In V^j, a mother function is first defined, and Haar's mother function is the simplest and was used by Balboa and Lopez (2000) for filtering.

Haar's mother function or analyzing wavelet is as follows:

$$\phi_i^j(x) = \phi(2^j x - i), \quad i = 0, 1, \cdots, 2^j - 1 \tag{6.17a}$$

$$\phi(x) = \begin{cases} 1 & if\ 0 \le 2^n x - i < 1 \\ 0 & in\ any\ other\ case \end{cases} \tag{6.17b}$$

The index j refers to *dilation,* and i refers to *translation*. The scale index j indicates the wavelet's width, and the location index i gives its position. The mother functions are rescaled, or "dilated," by powers of two and translated by integers.

Haar wavelets bases, like many others, are normalized by the constant $\sqrt{2^j}$, which is chosen such that

$$\left\langle \phi_i^j, \phi_i^j \right\rangle = \int_0^1 \phi_i^j(t)^2\, dt = 1$$

After the normalization, the wavelet coefficients of the example given in Figure 6.11 will become:

$$4, \quad -2, \quad \frac{1}{\sqrt{2}}, \quad \frac{1}{\sqrt{2}}, \quad -\frac{1}{2}, \quad \frac{1}{2}, \quad \frac{3}{2}, \quad 1$$

Filtering is then applied to this set of coefficients. For example, the smallest four coefficients are eliminated, and then the smoothed set is [4, −2, 0, 0, 0, 3/2, 1]. The reconstructed values become $f' = (2, 2, 2, 2, 3, 9, 8, 4)$. The effect is illustrated in Figure 6.12. The left panel is the original and the right panel is filtered. A practical example is given in Figure 6.13.

Figure 6.12 is an example produced by Balboa and Lopez (2000) using a real-life data set. It can be seen clearly that the low frequency part is so filtered out that the curve has become a straight line. The shape of this curve is quite distorted by such a filtering process. It might be not so appropriate to eliminate those small coefficients. Instead, we should think about the reduction of those big coefficients.

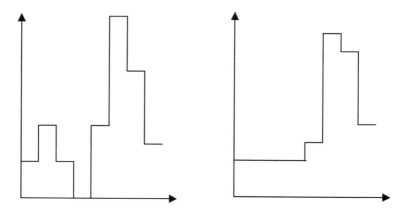

FIGURE 6.12 Effect of filtering on the wavelet coefficients (Modified from Balboa and Lopez, 2000. With permission.).

FIGURE 6.13 An example of filtering by wavelet using real-life data (Reprinted from Balboa and Lopez, 2000. With permission.).

6.5 SMOOTHING BY COMPONENT EXCLUSION IN THE SPACE DOMAIN

Empirical mode decomposition (EMD) is a new method of data analysis, designed for short-span data, which are nonlinear and nonstationary. EMD is a data-driven method. That is, the basis of decomposition is derived from the data instead of using an *a priori* basis (sine, cosine, or wavelet) as has been the case for many existing methods of data analysis, for example, Fourier and wavelet transforms (Huang et al., 1998). This adaptiveness has been found to be the most significant characteristic of EMD. EMD has been used for line smoothing by Li et al. (2004).

6.5.1 SMOOTHING BY EMD

The key idea of EMD is to decompose the data into a finite number of *intrinsic mode functions* (IMFs). An IMF is defined as any function that has the same number of zero crossings and extrema and that has symmetric envelopes defined by the local maxima and minima, respectively (Figure 6.14). In other words, an IMF can be considered to be a well-behaved signal with no riding or asymmetric waveforms.

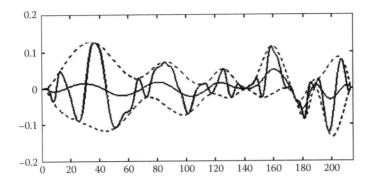

FIGURE 6.14 The envelope concept in EMD (Reprinted from Li et al., 2004).

To decompose a set of data into IMFs, one should first find the local extrema (i.e., maxima and minima) from the data. This can be done simply by finding slope changes at points on the line. In other words, any point with a change in the sign of slope (first derivative) is considered a local extrema. Having located these points, the next step is to fit spline curves onto them, that is, one curve for the maxima and the other for the minina. The former (i.e., the curve passing through the maxima) defines the upper envelope, and the latter (i.e., the curve passing through the minima) defines the lower envelope. Using the upper and lower envelopes, a set of mean values that define the mean envelope can be computed, as shown by the middle line in Figure 6.14.

By subtracting the mean envelope from the signal, the first IMF is then generated. The mathematical formula could be written as

$$h_1 = x(l) - m_1 \tag{6.18}$$

where $x(l)$ is the original data function, m_1 is the mean envelope, and h_1 is the first component.

Ideally, h_1 should be an IMF, but depending on the degree of nonstationarity of the data, new extrema might be generated in h_1. The process, therefore, can be iterated until the final result is completely an IMF. This process is called *sifting*, as it sifts the data to extract its sharp oscillations.

The standard deviation (σ) between the two successive h components can be used as a criterion for terminating the iteration. That is, the iteration stops if σ becomes smaller than a predefined threshold, for which a value of between 0.2 and 0.3 has been recommended by Huang et al. (1998). The threshold then becomes:

$$\sigma = \sum_{t=0}^{T} \left[\frac{\left|(h_{1(k-1)}(l) - h_{1k}(l))\right|^2}{h_{1(k-1)}^2(l)} \right] \leq 0.2 \tag{6.19}$$

In order to make sure that the resulting component is a complete IMF, a test to check the process is required. The test checks whether the number of zero crossings and extrema points are equal to, or differ by, one at most.

After this first component is subtracted from the data, the remainder r_1 can be treated as a new set of input data to the sifting process. That is,

$$r_1 = x(l) - c_1 \tag{6.20}$$

where, $c_1 = h_1$. The next component can then be extracted from the remainder r_1. By repeating this procedure several times, one can extract all of the components from the data as follows:

$$
\begin{aligned}
r_2 &= r_1 - c_2, \\
r_3 &= r_2 - c_3, \\
&\cdots, \\
r_n &= r_{n-1} - c_n
\end{aligned}
\tag{6.21}
$$

The process for extracting components is continued until the remainder becomes a monotonic function, from which no more IMFs can be extracted.

The reconstruction is to sum up the decomposed components as follows:

$$x(t) = \sum_{i=1}^{n} c_i + r_n \tag{6.22}$$

where $x(l)$ is the original data, c_i are the components, and r_n is the last component, which is a monotonic function showing the trend of the data. This equation indicates the completeness of the method because the original data can be reconstructed precisely as the components are summed (Figure 6.15). The two lines shown in Figures 6.15a and 6.15b should be identical. A possible difference between the original and reconstructed data would only be due to an error in rounding off the computations.

Figure 6.16 shows the sifting process. The curve as shown in Figure 6.15a is presented in parametric form, that is, by two functions, $X(l)$ and $Y(l)$. The sifting process is applied to these two functions separately, and the results are shown in Figures 6.16a and 6.16b.

Li et al. (2004) have used EMD for line smoothing. This idea comes from the fact that what has to be done for curve smoothing is a low-pass filtering, which is traditionally carried out in the domain of frequency. Both nonlinear and nonstationary data can be used to generate harmonic waves of all ranges. Therefore, any type of filtering in frequency space will eliminate some of the harmonics, which will cause deformations in the filtered data. By using IMF, however, one can also devise a time and space domain filtering.

A precise inspection of Figure 6.16 reveals that different components carry different frequency contents of data. One can also say the components are of different scales as compared to wavelet analysis. Such components are shown at the top of

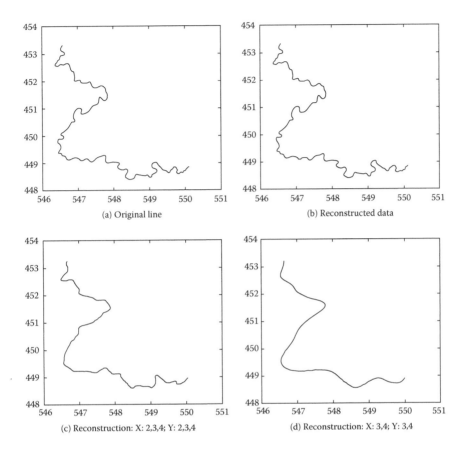

(a) Original line

(b) Reconstructed data

(c) Reconstruction: X: 2,3,4; Y: 2,3,4

(d) Reconstruction: X: 3,4; Y: 3,4

FIGURE 6.15 A line for testing and the reconstruction of a line by EMD (Reprinted from Li et al., 2004).

both Figures 6.16a and 6.16b. Therefore, the removal of such components from the data is equivalent to the omission of oscillations at a specific frequency. With an analogy to frequency analysis, a smoother version of the line will be generated by removing the first component of the data, which is of the highest frequency (and the smallest scale). Figure 6.15c shows the result. By the same logic, if a much smoother line is desired, the next components might be removed from the data. Figure 6.15d shows the result.

The procedure for line smoothing with EMD can be summarized as follows:

1. Separate the coordinates of a line into two sets, one set for X and the other set for Y.
2. Resample each set into equidistant points.
3. Decompose each set into a set of IMFs and the last monotone component.
4. Reconstruct the line with the exclusion of some high-frequency components.

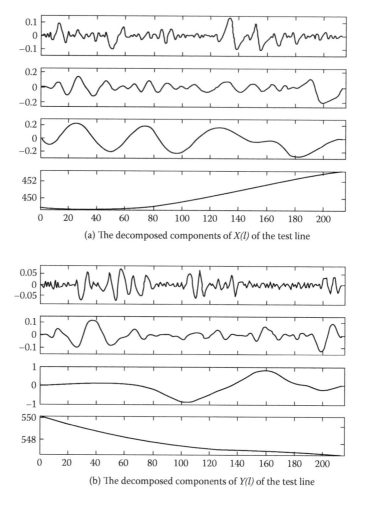

(a) The decomposed components of $X(l)$ of the test line

(b) The decomposed components of $Y(l)$ of the test line

FIGURE 6.16 The decomposition of the X and Y components of the line shown in Figure 6.15a (Reprinted from Li et al., 2004).

6.5.2 A COMPARISON BETWEEN EMD AND FREQUENCY-BASED TRANSFORMS

The Fourier transform is based on the representation of data as a superposition of sine and cosine functions. Therefore, the linearity of the system and the stationarity of the data are essential conditions for the Fourier analysis to be valid. However, the data generated from physical sources are often nonlinear and nonstationary. The Fourier transform produces a large number of harmonics to simulate such data. This leads to unwanted spurious energy and artificial frequencies. As a result, energy spreads over a wide range of frequencies. Constrained by the conservation of energy principle, these spurious harmonics and wide-frequency spectrum cannot faithfully represent the true density of energy in the frequency and time domain (Huang et al., 1998).

Wavelet is basically a windowing technique with variably sized regions that breaks up a signal into shifted and scaled versions of a mother wavelet. Hence, the wavelet spectrum is a plot of wavelet coefficients (amplitude) over time and scale axes. In other words, it yields the energy as a function of time and scale (frequency). The wavelet analysis provides a uniform resolution for all scales as an advantage. However, limited by the size of the basic wavelet function, for small scales (high frequencies), it is uniformly poor in resolution. The interpretation of the wavelet spectrum is also counterintuitive in some cases. For example, the occurrence of a local event in the low-frequency range will affect the high-frequency part of the spectrum, because local events are associated with a small scale and, hence, with high frequencies.

EMD is a data-driven method in which frequency has a physical meaning. In contrast with wavelet analysis, in EMD different frequencies indicate different modes of oscillation of the signal.

However, EMD is not without problems. With EMD, the resolution of the high-frequency signals in a time series may be reduced since the envelopes are repeatedly used in the processes of the signal decompositions. Furthermore, it is more difficult to determine the exact reasonable number of IMFs in a time series by means of EMD alone. The confused phenomena for the near-frequency signals in a time series may occur in the decomposed signals by EMD.

6.6 EVALUATION OF LINE SMOOTHING ALGORITHMS

Mokhtarian and Macworth (1992) suggested a number of criteria for evaluating a method for curve representation and recognition, from which the following list may be related to the evaluation of line smoothing algorithms:

1. **Efficiency:** The computational complexity of the algorithm is low, so it is capable of working efficiently.
2. **Ease of implementation:** The implementation of a computer program requires the least time on programming and debugging.
3. **Invariance:** The resultant feature should be invariant under translation, rotation, and scaling.

Rosin (1992) has made a list of significance measures for curve lines, which may be used for evaluation of the smoothing results. Some of these measures are based on pointwise errors between the original curve and the resultant curve, some are based on curvature, and others based on changes in curvature:

1. Maximum error.
2. Average square error.
3. Maximum curvature.

4. Maximum curvature.
5. Maximum curvature difference.
6. Average curvature difference.
7. Number of curvature extrema.
8. Number of zeros of curvature.
9. Fractal dimension.

REFERENCES

Balboa, J. L. G. and Lopez, F. J. A., Frequency filtering of linear elements by means of wavelets: a method and example, *Cartographic Journal*, 37(1), 39–50, 2000.

Borkowski, A., Burghardt, D., and Meier, S., A fast snakes algorithm using the tangent angle function, *International Archives of Photogrammetry and Remote Sensing*, 32(Part 3-2W5), 644–650, 1999.

Boutoura, C., Line generalization using spectral techniques, *Cartographica*, 26(3&4), 33–48, 1989.

Brenner, N. M., Fast Fourier transform of externally stored data, *IEEE Transactions*, AU-17, 128–132, 1969.

Burghardt, D. and Meier, S., Cartographic displacement using the snakes concept, in *Semantic Modeling for the Acquisition of Topographic Information from Images and Maps*, Foerstner W. and Pluemer, L., Eds., Birkhaeuser-Verlag, Basel, Switzerland, 1997, pp. 59–71.

Cooley, J. and Tukey, J. W., An algorithm for the machine calculation of complex Fourier series, *Mathematics of Computing*, 19, 297–301, 1965.

Graps, A., An introduction to wavelets, *IEEE Computational Science and Engineering*, 2(2): 50–61, 1995.

Huang, N. E., Shen, Z., Long, S. R., Wu, M. C., Shih, H. H., Zheng, Q., Yen, N.-C., Tung, C. C., and Liu, H. H., The empirical mode decomposition and the Hilbert spectrum for nonlinear and non-stationary time series analysis, *Proceedings of the Royal Society of London*, A454, 903–995, 1998.

Kass, M., Wikin, A., and Terzopoulos, D., Snakes: Active contour models, *International Conference on Computer Vision*, 1(4): 321–331, 1987.

Li, Z. L. and Su, B., Algebraic models for feature displacement in the generalization of digital map data using morphological techniques, *Cartographica*, 32(3), 39–56, 1996.

Li, Z. L. and Su, B., Some basic mathematical models for feature displacement in digital map generalization, *Proceedings of ICC'97*, 1, 452–459, 1997.

Li, Z. L., Khoshelham, K., Ding, X., and Zheng, D., Empirical mode decomposition (EMD) transform for spatial analysis, in *Advances in Spatial Analysis and Decision Making*, Li, Z., Zhou, Q., and Kainz, W., Eds., A.A. Balkema, Lisse, The Netherlands, 2004, pp. 19–30.

Mokhtarian, F. and Macworth, A. K., A theory of multiscale, curvature-based shape representation for planar curves, *IEEE Transactions on Pattern Analysis and Machine Intelligence*, 14(8), 789–805, 1992.

Plazanet, C., Affholder, J. G., and Fritsch, E. The importance of geometric modeling in linear feature generalization. *Cartography and Geographic Information Systems*, 22(4): 291–305, 1995.

Rosin, P., Representing curves at their natural scales, *Pattern Recognition*, 25(11), 1315–1325, 1992.

Steiniger, S. and Meier, S., Snakes: a technique for line smoothing and displacement in map generalization, in *ICA Workshop on Generalization and Multiple Representation*, August 20–21, 2004, Leicester, UK.

Strang, G., Wavelets, *American Scientist*, 82, 250–255, 1992.

Tobler, W., Numeric map generalization, Michigan Inter-University Community of Mathematical geographers, Discussion Paper No. 8, Reprinted in *Cartographica*, 26(1), 9–25, 1966.

Wang, Z. S. and Müller, J. C., Complex coastline generalisation, *Cartography and Geographic Information Systems*, 20(2), 96–106, 1992.

Weibel, R., A typology of constraints to line simplification, *Proceedings of 7th International Symposium on Spatial Data Handling (SDH '96)*, Springer, Berlin, 1996, pp. 9A.1–9A.14.

Weibel, R., Generalization of spatial data: principles and selected algorithms, in *Algorithmic Foundations of Geographic Information Systems,* Kreveld, M., Nievergelt, J., Roos, T., and Widmayer, P., Eds., Springer, New York, 2000, pp. 99–152.

Weisstein, E. W., *Cubic Spline*. MathWorld: A Wolfram Web Resource, http://mathworld. wolfram.com/CubicSpline.html, 2005.

Williams, D. and Shah, M., A fast algorithm for active contours and curvature estimation, in *Proceedings of the Third International Conference on Computer Vision*, IEEE Computer Society Press, Los Alamitos, CA, 1990, pp. 592–595.

Young, I. T., Walker, J. E., and Bowie, J. E., An analysis technique for biological shape, *Information and Control,* 25, 357–370, 1974.

(a) Topographic map at 1:200,000   (b) Thematic map (house development)

COLOR FIGURE 1.2 A topographic map (a) and a thematic map (b) (Courtesy of LIC of HKSAR).

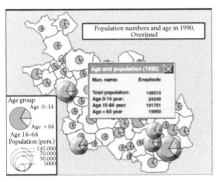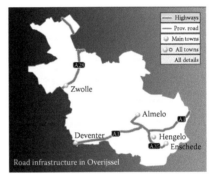

(a) Mouse-over symbol   (b) Legend-controlled display

COLOR FIGURE 1.6 New developments in interactive representations (van den Worm, 2001).

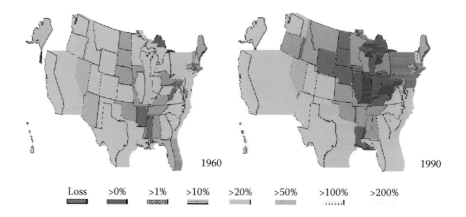

Loss >0% >1% >10% >20% >50% >100% >200%

COLOR FIGURE 1.8 Two frames selected from an animation of U.S. population change since 1790.

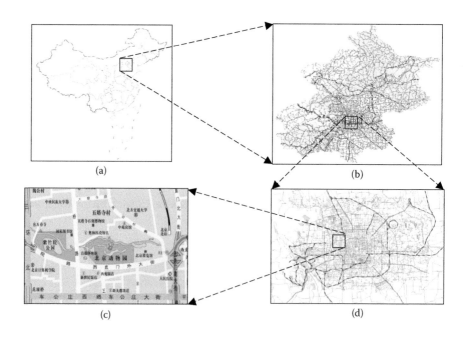

(a) (b) (c) (d)

COLOR FIGURE 1.9 Zooming into Beijing streets as an animation (Courtesy of National Geometics Center of China).

COLOR FIGURE 1.11 Changes of coastal lines in Hong Kong over time (Courtesy of LIC of HKSAR).

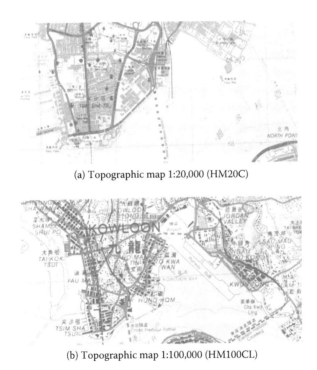

(a) Topographic map 1:20,000 (HM20C)

(b) Topographic map 1:100,000 (HM100CL)

COLOR FIGURE 1.12 The Kowloon Peninsula represented on maps at two different scales (Courtesy of LIC of HKSAR).

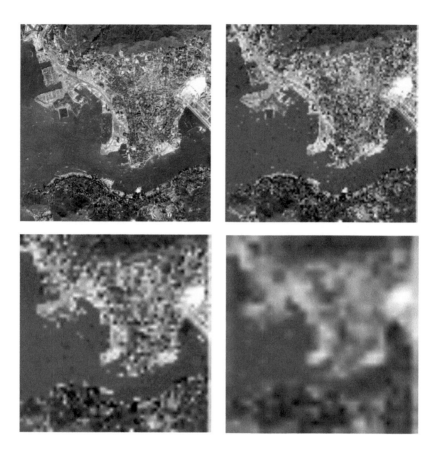

COLOR FIGURE 3.2 Four images with the same cartographic ratio but different resolutions.

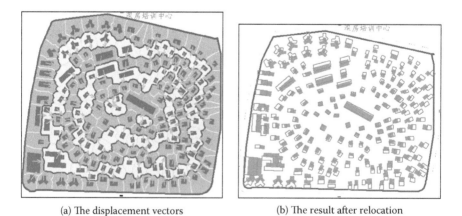

(a) The displacement vectors

(b) The result after relocation

COLOR FIGURE 11.14 Displacement field applied to building relocation caused by street widening (Reprinted from Ai, 2004. With permission.).

7 Algorithms for Scale-Driven Generalization of Individual Line Features

The previous two chapters introduced algorithms for point-reduction and smoothing of individual line features. In this chapter, another group of algorithms, for *scale-driven line generalization*, will be introduced. Scale-driven line generalization is a special type of line smoothing, which is directly related to the concept of scale.

7.1 SCALE-DRIVEN GENERALIZATION: AN OVERVIEW

As discussed in Chapter 5, in a traditional sense, generalization of a line means the simplification of the line structure to suit the representation at a smaller scale. That is, the main structure (characteristics) is retained while small details are removed. Figure 7.1 illustrates the concept of line simplification introduced by Keates (1989). Figure 7.1a shows the original line at the largest scale, and Figure 7.1b is generalized for representation at a smaller scale, that is, by a two times scale reduction. If the scale is further reduced by another two times and four times, the representations will be similar to the ones shown in Figure 7.1c and Figure 7.1d. The figure clearly shows the retention of the trend of the line.

However, Li (1993) has noticed some confusion and misunderstanding about the term *line simplification*. That is, many researchers have used this term to refer to the reduction of points on a line. In order to avoid such confusion, we use the term *scale-driven generalization* here to refer to the simplification illustrated in Figure 7.1.

It is clear that the traditional concept of simplification means to retain the main structure of a line but to remove the unwanted small details. Li (1993) has discussed the difference between point-reduction and generalization. Conceptually, point reduction is related to data compression and is used for saving storage and increasing efficiency of data processing. The number of points is of primary concern. Therefore, it is solely a digital processing problem, and there is no such a problem in conventional cartography. However, generalization, the purpose of which is to modify the original feature to suit the representation at a smaller scale while retaining the main characteristics of the original feature, always exists in both conventional and digital representation. Here the main characteristics of the original

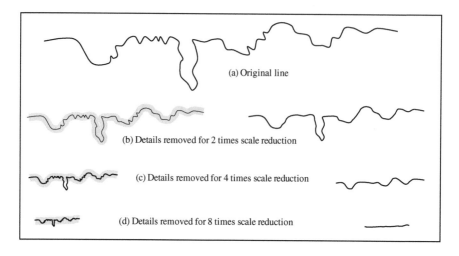

FIGURE 7.1 Traditional line simplification: retention of main structure.

line are of primary importance, but the number of points is not a concern at all. This view is shared by many researchers (e.g., Christensen, 1999, 2000). Thus, point reduction is applied only when scale change is not a concern, while generalization is applied only when there is a change in scale. If one is really concerned with the number of points, a data reduction process could be applied to the result obtained by scale-driven generalization.

An examination of existing algorithms published in the cartographic and geospatial literature reveals that only two algorithms belong to the category of scale-driven generalization. One is the classic ε-algorithm proposed by Perkal (1966), and the other is the algorithm by Li and Openshaw (1992a), based on the natural principle formulated by Li and Openshaw (1993).

It can also be found that *scale-space* becomes a concept in computing. Scale-space filtering has become popular for smoothing purposes. In this process one parameter is introduced to represent spatial scale and thus multi-scale representation is produced. Therefore, the author feels obliged to include some discussions about scale-space filtering, although the scale parameter in such techniques has no direct relationship to map scales.

7.2 ALGORITHMS BASED ON GAUSSIAN SPATIAL-SCALE

The scale-space approach has become very popular in multi-scale representation in computing since the publication of the seminal work by Witkin (1983). The concept of scale-space is an analogy to frequency space. The basic idea is to transform a curved line (or an image) by convolution with a Gaussian function into a new function in a new domain, which is called scale-space. The σ parameter in the Gaussian function (see Equation 7.2) is sometimes called the (Gaussian) *spatial-scale*. In this section, algorithms or procedures for smoothing based on Gaussian spatial scale will be described and the attempts to improve such algorithms will be discussed.

7.2.1 GAUSSIAN LINE SMOOTHING IN SCALE-SPACE

The basic idea of line smoothing in scale-space is to smooth the line as a signal by convolution with a one-dimensional (1-D) Gaussian function.

$$f(x,\sigma) = f(x) * G(x,\sigma) \tag{7.1}$$

where, * means the convolution operation, $f(x)$ is the original line signal, and $G(x, \sigma)$ is the Gaussian function with a standard deviation σ as follows:

$$G(x,\sigma) = \frac{1}{\sigma\sqrt{2\pi}} e^{\frac{-x^2}{2\sigma^2}} \tag{7.2}$$

Then, $f(x, \sigma)$ can be written as

$$f(x,\sigma) = \int_{-\infty}^{\infty} f(t) \times \frac{1}{\sigma\sqrt{2\pi}} e^{\frac{-(x-t)^2}{2\sigma^2}} dt \tag{7.3}$$

It should be noted here that Equations 7.2 and 7.3 are for the convolution of continuous functions. However, in digital representation of line features, the data are in discrete form. Therefore, discrete convolution should be used as follows:

$$y(x) = \sum_{t=1}^{K} f(t) \times h(x - t) \tag{7.4}$$

where K is the number of elements in $h(x)$.

If the number of points in $f(x)$ is N, then the result, $y(x)$, will have $(N + K - 1)$ points. Suppose $f(x) = \{9, 5, 3, 6, 14, 6, 9\}$ and $h(x) = \{1, 2, 3\}$; then the discrete convolution of these two digital representations in one dimension, $f(x) * h(x)$, is illustrated in Table 7.1. The first step is to flip the $h(x)$ into $h(-x) = \{3, 2, 1\}$ and then move the $h(-x)$ as a template along the $f(x)$. There are nine (= 7 + 3 − 1) points in the result: $\{9, 23, 40, 27, 35, 70, 105, 36, 27\}$.

In the case of smoothing by convolution, the degree of smoothing is controlled by σ. If $\sigma = 0$, the Gaussian function is a sharp pulse, leading to the result of $f(x,\sigma) = f(x)$, that is, no smoothing at all. With an increase in the value of σ, the Gaussian function becomes flatter and flatter (see Figure 7.2), and thus the weights become closer to equality, leading to a larger and larger smoothing effect. At $\sigma = \infty$, the scale-space signal is a flat line.

TABLE 7.1

Discrete Convolution of Two Digital Representations in One Dimension

f(x)	0	0	9	5	3	6	14	6	9	0	0		Results
Operation	× +	× +	× +	× +	× +	× +	× +	× +	× +	× +	×		
h(1-t)	3	2	1										9
h(2-t)		3	2	1									23
h(3-t)			3	2	1								40
h(4-t)				3	2	1						=	27
h(5-t)					3	2	1						35
h(6-t)						3	2	1					70
h(7-t)							3	2	1				105
h(8-t)								3	2	1			36
h(9-t)									3	2	1		27

In the case of smoothing by convolution with the Gaussian function, the operation has two special characteristics:

1. The Gaussian kernel is a symmetric template and thus there is a no need to flip the template.
2. The Gaussian kernel is usually normalized so that the sum of values within the template equals one.

The procedure of using scale-space smoothing can be written as follows:

1. Separate the coordinates of the line into two sets, one set for X, $X(l)$, and the other set for Y, $Y(l)$.
2. Resample each set into equidistant points.
3. Select a σ value for the Gaussian function $G(l, \sigma)$.
4. Convolve $X(l)$ and $Y(l)$ with the selected Gaussian function $G(l, \sigma)$, resulting in $X(l, \sigma)$ and $Y(l, \sigma)$.
5. Obtain the smoothed line from the smoothed X and Y sets, that is $X(l, \sigma)$ and $Y(l, \sigma)$.

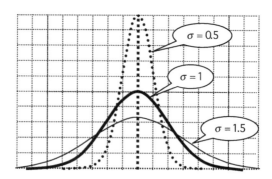

FIGURE 7.2 A normal distribution with different σ values.

It is clear by now that smoothing by convolution with a Gaussian function is exactly the same as smoothing by moving averaging with a Gaussian kernel as weighting. In this sense the materials described in this section should be included in Section 6.2. However, the emphases are different. In this section the main idea is the control of the smoothing effect with σ as a scaling variable, and thus multi-scale representation can be produced by varying the σ value, sometimes termed *spatial-scale* in the literature.

Although Gaussian filtering is the most popular linear smoothing technique, there are problems associated with it. The most serious problem in multi-scale spatial representation is that the σ variable is not directly related to the scales of the representation, for example, map scales; thus, this technique as described in this section is not a scale-driven generalization as such.

7.2.2 ATTEMPTS TO IMPROVE GAUSSIAN SMOOTHING

In the computing community there is a concern about the shrinkage problem created by Gaussian smoothing, and efforts have been made to solve the problem.

Shrinkage refers to the phenomenon where the smoothed curved line is systematically shrunk toward the center of curvature. For example, a smaller circle will be produced after it is smoothed. This is a problem associated with all averaging filters.

Horn and Weldon (1986) have addressed the problem for closed convex curves. Their method is based on the extended circular image, in which the radius of curvature is given as a function of normal directions. More precisely, in the convolution process, they made use of the radius of curvature (instead of curvature itself) and used the normal angle (instead of arc-length) as an independent variable. However, this method is apparently limited to closed convex curves (Rosin, 1992; Oliensis, 1993).

Lowe (1989) has tried to solve the shrinkage problem. His method works as follows:

1. Use the standard Gaussian smoothing to smooth the curved line.
2. Estimate the amount of shrinkage at each point along the line as a function of curvature.
3. Inflate the curve to correct for the estimated shrinkage.

The shrinkage at a point is computed from the following:

$$S(l) = r\left(1 - e^{\frac{-\sigma^2}{2r^2}} \right) \tag{7.5}$$

where r is a parameter given as follows:

$$X''(l) = e^{\frac{-\sigma^2}{2r^2}} \Big/ r \tag{7.6}$$

where $X''(l)$ is the convolution of $X(l)$ with $G''(l, \sigma)$.

Equation 7.6 needs to be solved numerically for r. The shrinkage correction is usually implemented in a look-up table form.

Oliensis (1993) found that the shrinkage in Gaussian smoothing is simply due to the reduction in the low- as well as the high-frequency components and implemented a local reproducible smoothing method as follows:

1. Transform the coordinate functions $X(l)$ and $Y(l)$ using fast Fourier transform (FFT) to yield frequency representation $X(k)$ and $Y(k)$.
2. Cut off the frequency at $|k| > 1/\sigma$.
3. Transform the new frequency representation back to the space domain by inverse FFT.

7.3 ALGORITHMS BASED ON ε-CIRCLE ROLLING

Perkal's attempt to produce a scale-driven objective line generalization (Perkal, 1966) was the first effort of its kind. In this section the principles of Perkal's original algorithm are described and other attempts to make the Perkal algorithm more useable are presented.

7.3.1 PERKAL ALGORITHM BASED ON ε-CIRCLE ROLLING

The *Perkal algorithm* is centered on the assignment of a scale-dependent weight to the lines to be generalized. As pointed out by Christensen (2000), "it is due to such a dependency, Perkal algorithm stands apart from the rest of the field of so-called generalization algorithms that essentially do little more than to skip points."

The principle of the Perkal ε-*circle rolling algorithm* is to roll a circle with a diameter of ε on both sides of a line so that the small details can be removed. Figure 7.3 shows the process. The general idea of the Perkal ε-circle rolling algorithm is to produce a curved line with curvature larger than a certain value with guarantee. The curvature

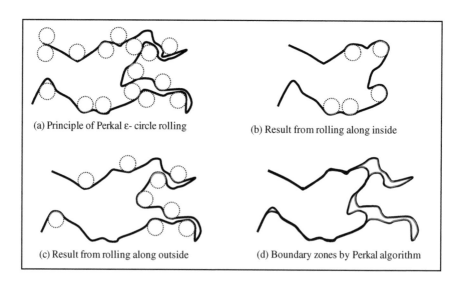

(a) Principle of Perkal ε- circle rolling

(b) Result from rolling along inside

(c) Result from rolling along outside

(d) Boundary zones by Perkal algorithm

FIGURE 7.3 Objective generalization of a curved line by Perkal's ε-circle rolling algorithm.

is determined by the size of the circle used for rolling. The diameter of the circle is ε, and thus the radius is $\delta = \varepsilon/2$. The smallest curvature of the resultant line is then equal to $C_{min} = 1/\delta = 2/s$. The value of ε is determined by the scale of the target map (e.g., 1:100,000) and the line weight on map (e.g., 0.2 mm).

The algorithm can be written as follows:

1. Determine a value for the ε-circle.
2. Roll the ε-circle along each side of the line to replace the sharp bends with the arcs of the ε-circle.
3. Overlay the two results to obtain a generalized line.

In Figure 7.3, some ε-circles are placed on those sharp bends of the line, meaning that such bends are too sharp to be retained at a smaller scale, and the corresponding circle arcs should be used to replace these bends. Figures 7.3b and 7.3c show that two results are obtained, one from the rolling of the circle on each side of the line. Figure 7.3d shows the final result, which is the overlay of these two results shown in Figures 7.3b and 7.3c. This shadowed area is called the *generalized edge* or *boundary zone* (see Christensen, 2000).

The original proposal by Perkal (1966) ends with such a definition of generalized edge but did not propose any method for further processing. As has been pointed out by many (e.g., Brophy, 1973; Zoraster et al., 1984; Beard, 1991), Perkal in his original work only applied the procedure manually, and computerization of the Perkal algorithm is not easy. Christensen (2000) also comments that "in his time and in terms of manual generalization, that definition was all that was needed to meet his objective: a method for letting the cartographers know which section of a line should be generalized." However, it is not a complete story, and the attempts to improve the Perkal algorithm will be reported in Sections 7.3.2 and 7.3.3.

7.3.2 The **WHIRLPOOL** Approximation of the Perkal Algorithm

The Perkal algorithm assures an ε-convexity. That is, every point on the resultant line will have a radius of curvature greater than or equal to ε. A close approximation to the Perkal algorithm is the *WHIRLPOOL* implementation (Dougenik, 1980), which also assures ε-convexity for the resultant line.

In the WHIRLPOOL algorithm a distance criterion d is used. This is not the perpendicular distance used in the curve approximation algorithms discussed in Chapter 5. It is similar to the ε concept in the Perkal algorithm. In the WHIRLPOOL algorithm all the points are checked to see whether the distance between any two points is smaller than d. In the end, those points that are closer than the given criterion d are removed from the point clusters. This ensures that the resultant line will have points at least d apart. Figure 7.4 illustrates this concept. The result is very desirable in this case. However, the result might be dependent on the interval between points on the line. It is possible to leave out some points in the spiky area if the interval between the points in this area is larger than d.

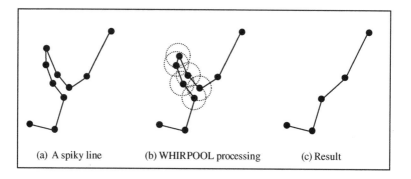

FIGURE 7.4 The working principle of the WHIRPOOL algorithm.

7.3.3 WATERLINING AND MEDIAL AXIS TRANSFORMATION FOR PERKAL'S BOUNDARY ZONE

Christensen (2000) identified the boundary zone problem associated with the Perkal algorithm. He points out that "a computer generalization requires a further step: a way to connect the ε-convex lines across the boundary zones." He proposed to apply medial axis transformation to derive a line across the area. Figure 7.5 illustrates this idea.

The line derived by medial axis transformation is a skeleton of the boundary area. There are also other methods for the derivation of skeletons (e.g., Bookstein, 1979). A detailed description of skeletonization algorithms has been given in Chapter 2.

Christensen (1999) also noticed that the buffering operation in modern geographical information systems and the waterlining operation used in art produce representations similar to the area formed by the locus of the ε-circle rolling inside and outside the line. The medial axis transformation can be applied to derive the central line, which can then be regarded as the generalized results.

The medial axis of a boundary zone with a simple area shape can be easily derived and can be used to represent the generalized line across the area. However, the suitability of using such middle lines as the generalized lines for the two spiky boundary zones as shown in Figure 7.3d is questionable. Therefore, the effectiveness of Christensen's solution needs to be evaluated.

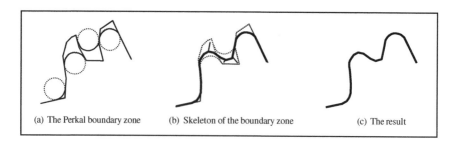

FIGURE 7.5 Derivation of a skeleton in the boundary zones generated by the Perkal algorithm.

7.4 ALGORITHMS BASED ON THE NATURAL PRINCIPLE

The idea introduced by Perkal (1966) is very interesting, although his algorithm does not work in practice. Inspired by Perkal's idea, Li and Openshaw (1992a) developed a set of algorithms based on the nature principle of objective generalization (Li and Openshaw, 1993) described in Chapter 3. They developed three algorithms: vector mode, raster mode, and raster–vector mode algorithms. All of them are scale driven, that is map scales are the only parameters in the algorithms. In their original paper Li and Openshaw (1992a) recommended the algorithm implemented in raster–vector mode. Weibel (1997) called this algorithm the Li–Openshaw algorithm. This section describes the algorithms implemented in raster and raster–vector modes. The implementation of the Li–Openshaw algorithm for a set of lines will be discussed in Chapter 8.

7.4.1 THE BASIC IDEA OF LI–OPENSHAW ALGORITHM

The basic idea employed in the Li–Openshaw algorithm is the natural principle. That is, at a given scale, an object (feature) with a physical size smaller than a certain limitation cannot be seen and thus can be completely neglected. Li and Openshaw (1992a, 1993) called the object (feature) with such a limitation is called the smallest visible object (SVO). This limitation may be better termed smallest visible size (SVS).

Figure 7.6 illustrates the neglect of all spatial variations of a complex line feature within an SVS so as to generalize the line. Figure 7.6a shows a complex line and SVS. However, no matter how complex the line is, all spatial variations within the SVS are completely neglected, as shown in Figure 7.6b, and the SVS is represented by a point in vector or by a pixel (or the center of the pixel) in raster, as shown in Figure 7.6c.

The computation of the SVS is expressed by Equations 3.2 and 3.3, which are dependent on the scales of the source map and target map.

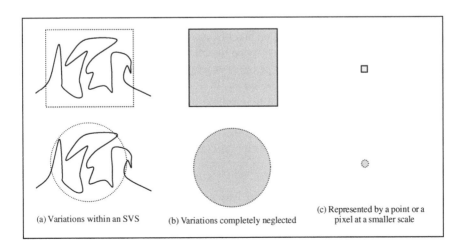

(a) Variations within an SVS

(b) Variations completely neglected

(c) Represented by a point or a pixel at a smaller scale

FIGURE 7.6 A point or a raster cell can be used to represent the spatial variations within an SVS.

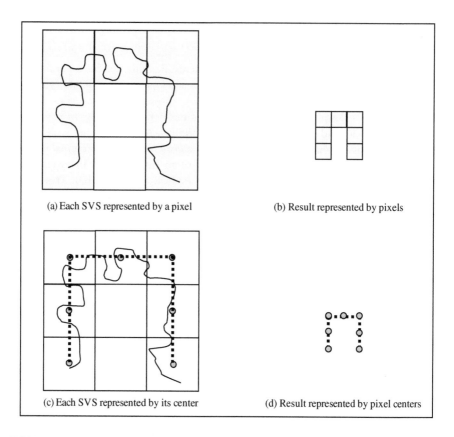

(a) Each SVS represented by a pixel

(b) Result represented by pixels

(c) Each SVS represented by its center

(d) Result represented by pixel centers

FIGURE 7.7 Li–Openshaw algorithm in raster mode, with each cell being an SVS.

7.4.2 THE LI–OPENSHAW ALGORITHM IN RASTER MODE

In this implementation a cell or pixel is used to mimic the SVS. The cells can be organized in a form of nonoverlapped tessellation or in a form with overlaps. If there is no overlap, it becomes a pure raster template. Figure 7.7 shows the generalization process with the raster template. In Figure 7.7a each SVS is represented by a raster pixel, and the result is represented by pixels, as shown in Figure 7.7b. In Figure 7.7c each SVS is represented by its center, and the result is represented by SVS centers, as shown in Figure 7.7d.

This algorithm for the generalization of individual lines can be implemented by moving the cell as follows:

1. Determine a value for the SVS based on Equations 3.2 and 3.3.
2. Center the first SVS cell at the first point of the curved line and record the first cell (or point).
3. Compute the intersection between the curved line and the boundary of the cell to determine the exit direction of the line in the current cell.

4. Place a cell in the exit direction. This cell becomes the current cell.
5. Record this cell and repeat steps 3 and 4 until all points have been considered.

The cells can be moved along four directions (in the case of four-adjacency) or eight directions (in the case of eight-adjacency). This algorithm can also be implemented by a process similar to rasterization, as follows:

1. Determine a value for the SVS based on Equations 3.2 and 3.3.
2. Put a raster template arbitrarily onto the curved line, with the raster size being equal to the SVS.
3. Locate each point of the curved line in the raster template and record the located cells.
4. Remove the duplicated cells recorded, if any.

As will be discussed in Chapter 8, this rasterization approach is more desirable if a set of lines, for example, contour lines, is to be generalized. Figure 7.8 shows an example of generalization of the closed line (Figure 7.8a), using this approach for a 20 times scale reduction. It is clear that there are some undesirable parasitic branches (Figure 7.8b), and a pruning process is needed. The final smoothed result is shown in Figure 7.8c.

The cells can also be overlapped for two purposes. The first is to remove the dependency of results on the starting point, that is, where the first cell is placed. The second purpose is to retain more details. Figure 7.9 illustrates the process with overlapping cells, using the moving SVS approach. The center of the first cell is placed near the starting point (Figure 7.9a). Then the cells are moved along the line direction. When the line turns, the cell direction is changed accordingly, as shown in Figure 7.9b. The complete process is shown in Figure 7.9c. In this example, only four directions are considered. It is also possible, and indeed even better, to consider eight directions in practice, as shown in Figure 7.9d. Li and Openshaw (1992a) experimented with that effect of the overlapping percentage being not very significant. A value of 20% is sufficient, and zero overlap is still acceptable in practice.

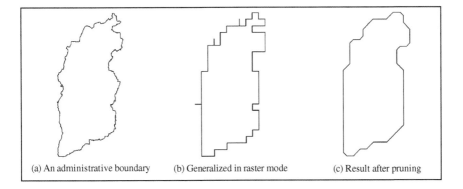

(a) An administrative boundary   (b) Generalized in raster mode   (c) Result after pruning

FIGURE 7.8 Generalization of a real-life administrative boundary in raster mode.

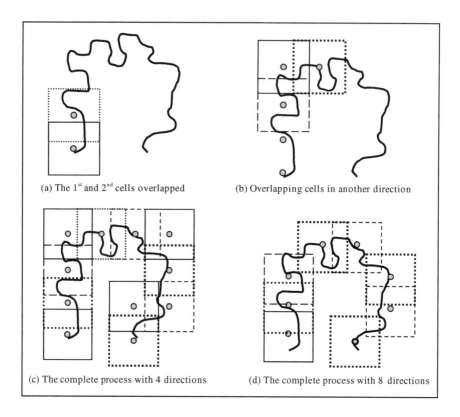

(a) The 1st and 2nd cells overlapped

(b) Overlapping cells in another direction

(c) The complete process with 4 directions

(d) The complete process with 8 directions

FIGURE 7.9 Generalization with overlapping cells, using the moving cell approach.

7.4.3 THE LI–OPENSHAW ALGORITHM IN RASTER–VECTOR MODE

As discussed in the previous subsection, the overlap percentage is not critical, so the implementation with a simple raster template is quite acceptable. Such an implementation without overlap is computationally simple. It has also been noticed that the results from using the algorithm in raster mode are not very smooth visually. Therefore, Li and Openshaw (1992a) recommended a raster–vector implementation, which means the generalization process is in raster mode but the representation of the result is in vector mode.

The implementation should be very similar to the algorithm in raster mode because the principle is the same. In other words, both approaches (i.e., rasterization and moving cell) can be used. The major difference is the representation of an SVS. In raster mode a whole pixel or the center of the pixel is used to represent an SVS. In vector mode any point within an SVS can be used to represent the SVS. Examples of such points could be:

1. The first point on the line but within the SVS.
2. The last point on the line but within the SVS.
3. Any point on the line but within the SVS.
4. The center of the SVS.

5. The average of all points on the line but within the SVS.
6. The average of the intersections between the line and the boundary of the SVS.
7. The middle of the first and last intersection between the line and the boundary of the SVS.

It is also possible to represent an SVS only by the intersections between the line and the SVS.

Figure 7.10 illustrates an example of the working principle of the Li–Openshaw algorithm in raster–vector mode. The first step is to lay down a raster template on the line to be generalized (Figure 7.10a). The first point to be recorded is the starting point (Figure 7.10b). The second point is somewhere within the second cell. This implementation uses the middle point between the two intersections between cell grids and the line (Figure 7.10b). If there is more than one intersection, the first (from the inlet direction) and the last (from the outlet direction) intersections are used to determine the position of the new point (Figure 7.10c). The final result of the generalization of a complete line is given in Figure 7.10d.

Figure 7.11 shows the generalization of an administrative boundary by the Li–Openshaw algorithm in raster–vector mode. Figure 7.11a shows the original line.

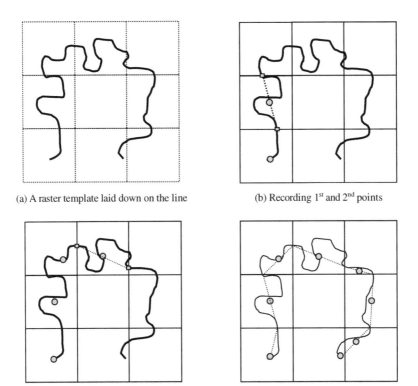

(a) A raster template laid down on the line   (b) Recording 1st and 2nd points

(c) If more than one intersection, take 1st and last   (d) Generalization of a complete line

FIGURE 7.10 The Li–Openshaw algorithm in raster–vector mode.

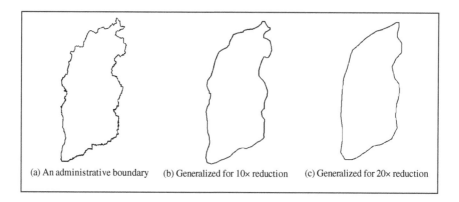

(a) An administrative boundary   (b) Generalized for 10× reduction   (c) Generalized for 20× reduction

FIGURE 7.11 Generalization of a real-life administrative boundary in raster–vector mode.

Figures 7.11b and 7.11c show the results of generalization using the raster–vector algorithm for scale reduction by 10 times and 20 times, respectively. It is clear that after generalization the main structure is retained, while the small variations are removed.

Similar to the algorithm in raster mode, overlap between SVSs could also be adopted, although this is not too critical. If there is no overlap, the generalization process is similar to simple resampling.

7.4.4 Special Treatments in the Li–Openshaw Algorithm

Figure 7.8 demonstrated that some parasitic branches will be generated for those narrow spiky lines if the algorithm in raster mode is applied. Thus, pruning is needed. However, with the algorithm in raster–vector mode, some of the narrow spiky features may be deleted because they are smaller than the SVS. Special attention needs to be paid to the thin necks of lines. Figure 7.12a shows an example of this. If the neck is too thin (thinner than two cells), there are three options:

1. Throw away the small convex parts (Figure 7.12b).
2. Form a close loop for the small convex parts (Figure 7.12c).
3. Exaggerate the thin necks (Figure 7.12d).

Some additional constraints must be imposed while exaggerating the concave parts. When only considering scale-driven generalization, experience shows that if the convex parts occupy fewer than four cells, then the first option is taken; otherwise, the second option is taken.

7.4.5 The Li–Openshaw Algorithm for Nonnatural Lines: Some Remarks

It is a misconception to say that the Li–Openshaw algorithm is only suitable for the generalization of natural boundaries. If the rasterization approach is adopted and the points on the original line are used, then the Li–Openshaw algorithm can also be used for the simplification of artificial lines, for example, cadastral boundaries.

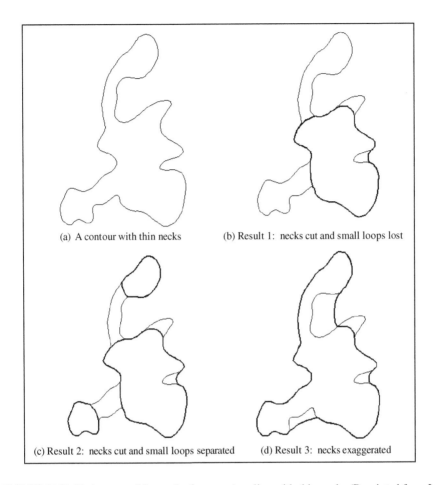

(a) A contour with thin necks (b) Result 1: necks cut and small loops lost

(c) Result 2: necks cut and small loops separated (d) Result 3: necks exaggerated

FIGURE 7.12 Various possible results for a contour line with thin necks (Reprinted from Li and Sui, 2000. With permission.).

Figure 7.13 shows two implementations of this. Figure 7.13a is an artificial line, and Figures 7.13b and 7.13c are the two results. In both cases the start and end points are first recorded, and the points on the original lines are retained if there is no more than one point in a cell. The only difference is the treatment when there is more than one point in a cell. In Figure 7.13b the (x,y) averages of all the points within a cell are used as the representative, while in Figure 7.13c no new points are created and the first point in a cell is taken as the representative.

7.5 EVALUATION OF SCALE-DRIVEN LINE GENERALIZATION ALGORITHMS

The previous sections described algorithms for the scale-driven generalization of curved lines. A question arising is, "How do we evaluate them?"

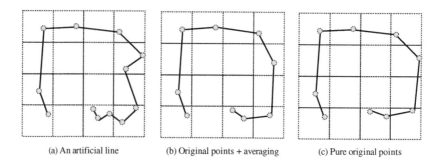

(a) An artificial line   (b) Original points + averaging   (c) Pure original points

FIGURE 7.13 Generalization of a nonnatural line in raster–vector mode.

7.5.1 Benchmarks for Evaluating Scale-Driven Line Generalization

There is not much literature addressing the performance of line generalization algorithms. Most research papers in this area are about the performance of point-reduction algorithms. However, as pointed out by Visvalingam and Whyatt (1990), the measures used for point-reduction are inappropriate, misleading, and questionable for scale-driven generalization.

The first issue to be tackled is the benchmarks to be used. In the case of point-reduction the original line is used as the benchmark. Is such a benchmark still applicable in the case of line generalization? Li and Openshaw (1992a,b) and Li (1993) consider that it is not appropriate to use the original line as a benchmark and suggested the use of a manually generalized result at target scale. For example, a line is generalized from 1:10,000 source maps to suit the representation at 1:50,000; the existing map at 1:50,000 should be used as a benchmark. Another possible benchmark is the results derived from aerial photographs at an appropriate smaller scale.

Li (1993) pointed out that if the appropriate benchmarks (as suggested here) are used, then the measures used for the evaluation of point reduction, such as vector and area displacements, and fractal dimension can be used without problems. Li and Openshaw (1992) found that Fourier spectrum can also be used for such a purpose.

Weibel (1996) has established a set of constraints for the manipulation of line features. These are mainly for point-reduction algorithms, but some of them are still valid for scale-driven line generalization. These constraints are as follows:

1. Avoid imperceptible crenulations.
2. Avoid self-coalescence.
3. Minimize shape distortion.
4. Avoid self-intersection.
5. Preserve the original line character.

7.5.2 Performance of Scale-Driven Line Generalization Algorithms

Not much evaluation work is available for the Perkal algorithm, simply because this algorithm does not provide a complete solution. It might be expected that this

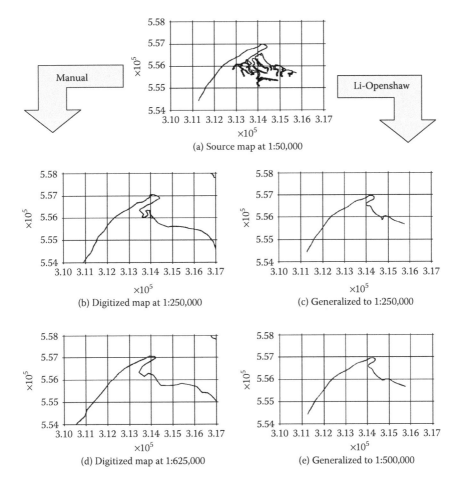

(a) Source map at 1:50,000

(b) Digitized map at 1:250,000

(c) Generalized to 1:250,000

(d) Digitized map at 1:625,000

(e) Generalized to 1:500,000

FIGURE 7.14 A comparison of the results of manual generalization and the Li–Openshaw algorithm (Reprinted from Li and Openshaw, 1992a.).

algorithm could still satisfy most of these constraints, because it is a scale-driven algorithm. As discussed previously, Gaussian smoothing is not scale-driven generalization as such, so the evaluation of this type of technique is not discussed here.

From the literature it can be found that self-intersection and cross-intersection are the most serious problems with many algorithms for line simplification. The Li–Openshaw algorithm, as pointed out by Weibel (1996), "by virtue of its raster structure, implicitly (but not explicitly) avoids self-overlaps." Even for a very complex coastline, it can still produce results extremely similar to manually generalized results. Figure 7.14 shows such a capacity of this algorithm, which is extracted from Li and Openshaw (1992a). Figure 7.14a is a coastline digitized from a 1:50,000 topographic map—original line in this figure. On the left side of the figure the two

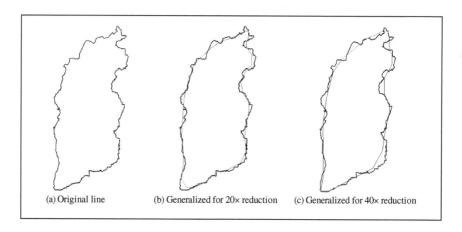

(a) Original line   (b) Generalized for 20× reduction   (c) Generalized for 40× reduction

FIGURE 7.15 Generalization of a medium complex administrative boundary by the Li–Openshaw algorithm.

lines were digitized from maps at smaller scales (1:250,000 and 1:625,000), which are supposed to be the result of manual generalization. On the right side the two lines at similar smaller scales (1:250,000 and 1:500,000) were generalized by the Li–Openshaw algorithm in raster–vector mode. The results clearly show that the Li–Openshaw algorithm is capable of producing results very similar to manual work, even when the line is extremely complex.

If distortion is concerned, the Li–Openshaw algorithm also performs extremely well. Figure 7.15 shows the generalization of a medium complex administrative boundary by the Li–Openshaw algorithm in raster–vector mode. Figure 7.15a shows the original line. Figure 7.15b shows the line generalized for 20 times reduction but enlarged back to the original scale and then superimposed onto the original map. Figure 7.15c shows the line generalized for 40 times reduction but enlarged back to the original scale and then superimposed onto the original map. From this figure, one can see that small details are removed from the original line while the main trend is retained.

REFERENCES

Beard, M. K., Theory of the cartographic line revisited: implications for automated generalization, *Cartographica*, 28(4), 32–58, 1991.

Bookstein, F.L., The line-skeleton, *Computer Graphics and Image Processing*, 11, 123–137, 1979.

Brophy, D. M., An automated methodology for linear generalization in thematic cartography, in *Proceedings of ACSM Conference 33rd Annual Meeting*, Washington, DC, 1973, pp. 300–314.

Christensen, A. H., Cartographic line generalization with waterlines and medial-axes, *Cartography and Geographic Information Science*, 26(1), 19–32, 1999.

Christensen, A. H., Line generalization by waterline and medial-axis transformation: success and issues in an implementation of Perkel's proposal, *Cartographic Journal*, 26(1), 19–32, 2000.

Dougenik, J. A., WHIRLPOOL: a program for polygon overlay, *Proceedings of Auto-Carto 4*, 1980, pp. 304–311.

Horn, B. K. P. and Weldon, E. H. Jr., Filtering closed curves, *IEEE Transactions on Pattern Analysis and Machine Intelligence,* 8(5), 665–668, 1986.

Keates, J., *Cartographic Design and Production*, 2nd ed., Longman Scientific, Essex, UK, 1989.

Li, Z. L., Some observations on the issue of line generalisation, *Cartographic Journal*, 30(1), 68–71, 1993.

Li, Z. L., An examination of algorithms for the detection of critical points on digital cartographic lines, *Cartographic Journal*, 32(2), 121–125, 1995.

Li, Z. L. and Openshaw, S., Algorithms for automated line generalisation based on a natural principle of objective generalisation, *International Journal of Geographic Information Systems,* 6(5), 373–389, 1992a.

Li, Z. L. and Openshaw, S., A comparative study of the performance of manual generalization and automated generalizations of line features, in *AGI 1992-1993 Year Book*, 1992b, pp. 502–513.

Li, Z. L. and Openshaw, S., A natural principle for objective generalisation of digital map data, *Cartography and Geographic Information Systems*, 20(1), 19–29, 1993.

Li, Z. L. and Sui, H. G., An integrated technique for automated generalisation of contour maps, *The Cartographic Journal,* 37(1): 29–37, 2000.

Lowe, D. G., Organization of smooth image curve curves at multiple scales, *International Journal of Computer Vision*, 3, 119–130, 1989.

Oliensis, J., Local reproducible smoothing without shrinkage, *IEEE Transactions on Pattern Analysis and Machine Intelligence*, 15(3), 307–312, 1993.

Perkal, J. D., *An Attempt at Objective Generalisation* (transl. Jackowski, W., from Proba obiektywnej generalizacji, *Geodezia I Kartografia*, Tom VII, Zeszyt 2, 130–142, 1958), Discussion Paper No. 10, Dept. of Geography, University of Michigan, Ann Arbor, 1966.

Rosin, P., Representing curves at their natural scales, *Pattern Recognition*, 25(11), 1315–1325, 1992.

Shapiro, B., Pisa, J., and Sklansky, J., Skeleton generation from x,y boundary sequences, *Computer Graphics and Image Processing*, 15, 136–153, 1981.

Visvalingham, M. and Whyatt, J., The Douglas-Peucker algorithm for line simplification: Reevaluation through Visualization, *Computer Graphics Forum,* 9:213–228, 1990.

Weibel, R., A typology of constraints to line simplification, in *Proceedings of 7th International Symposium on Spatial Data Handling (SDH '96)*, 1996, pp. 9A.1–9A.14.

Weibel, R., Generalization of spatial data: principles and selected algorithms, in *Algorithmic Foundation of Geographic Information Systems*, van Kreveld, M., Nievergelt, J., Roos, T., and Widmayer, P., Eds., Springer, Berlin, 1997, pp. 99–152.

Witkin, A. P., Scale-space filtering, in *Proceedings of the 8th International Joint Conference on Artificial Intelligence*, 1983, pp. 1019–1022.

Zoraster, S., David, D., and Hugus, M., *Manual and Automated Line Generalization and Feature Displacement,* ETL-0359, Army Engineering Topographic Laboratories, Fort Belvoir, VA, 1984.

8 Algorithms for Transformations of a Set of Line Features

From Chapter 5 to Chapter 7, various types of algorithms for individual line features have been discussed. This chapter will present algorithms for a set of lines.

8.1 A SET OF LINE FEATURES: AN OVERVIEW

On a spatial representation, there are many types of line features, such as boundary lines, coastlines, river networks, transportation networks, and contours. The boundary of an area feature (a closed line feature) will be treated in Chapter 9. A coastal line may form a closed line as the boundary of an area (i.e., an island) and may be treated as an area boundary. If it is an open line, the treatments presented in Chapters 5 to 7 can be applied. In this chapter only the multi-scale representation of contour lines, river networks, and transportation networks will be discussed.

A set of rivers form a river system. Each river system has an intrinsic hierarchical (tree) structure, as shown in Figure 8.1a. For a river network, two operations are typically applied: selective omission and scale-driven generalization. The former eliminates the less important branches, and the latter makes the variation of the selected rivers simpler to suit representations at a smaller scale. As the algorithms for the latter have been described in the previous chapters, only the selective operations will be addressed in this chapter. To make proper selective omission, researchers have made use of various ordering schemes for such tree structures developed by geomorphologists. One section of this chapter is devoted to the use of such ordering systems for generalization of river networks.

Roads (and railways) together form a network structure, rather than a tree structure, in a physical sense. However, some roads are more important than others, so there is also a hierarchy structure in a thematic sense. The information about the planar network and road hierarchy can be used for the selective omission of road segments, which is a topic of Section 8.3.

Contour lines have been used for the representation of terrain surfaces and other three-dimensional (3-D) surfaces since the 18th century. Contouring has been widely regarded as the most effective means for the representation of 3-D variations on a

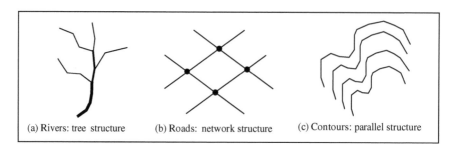

(a) Rivers: tree structure   (b) Roads: network structure   (c) Contours: parallel structure

FIGURE 8.1 Three types of line features.

2-D plane and has therefore been considered one of the most important inventions in history of spatial representation. Section 8.2 is devoted to the generalization of contour lines.

8.2 ALGORITHMS FOR TRANSFORMATION OF A SET OF CONTOUR LINES*

8.2.1 APPROACHES TO THE TRANSFORMATION OF CONTOUR LINES

The transformations for the multi-scale representation of a set of contour lines can be performed via two approaches: direct and indirect. The direct approach, as the name implies, works on the contour lines directly. The indirect approach produces the multi-scale representation via a digital terrain model (DTM) or digital elevation model (DEM).

The algorithm based on the direct approach is very simple:

1. Select a subset of contour lines from the original set.
2. Perform the transformation of the contour lines.

The algorithm based on the indirect approach takes three steps (Chen, 1987):

1. Generate a DTM or DEM from the contour lines.
2. Perform the transformation of the DTM.
3. Produce a new set of contour lines from the transformed DTM.

In this indirect approach steps 1 and 3 lie outside the scope of this book, but detailed discussion of the production of DTMs from contour data an be found in the recent book by Li et al. (2005). Step 2 will be discussed in Chapter 12. In this chapter, only the direct approach will be discussed.

---

* This section is largely extracted from Li and Sui, 2000. With permission.

8.2.2 SELECTION OF A SUBSET FROM THE ORIGINAL SET OF CONTOUR LINES: SELECTIVE OMISSION

To produce the representation of contour lines for a smaller scale, the following requirements must be fulfilled:

1. The structure of the contour lines must be simplified and smoothed.
2. The natural characteristics of contour lines (e.g., being parallel, without self- and cross-intersections, geometrically similar to the shape of the 3-D surface) must be retained.
3. No coalescence should be created.

Requirements 1 and 2 should be met automatically by generalization algorithms, but requirement 3 is usually met through the selective omission of contour lines.

When the scale change is dramatic, then the spacing between two contour lines (i.e., the planimetric contour interval) will be reduced to such a level that contour lines will be coalesced or touch each other. As a result, it is necessary to remove some of the contour lines to retain the clarity of the maps, that is, to avoid the coalescence, as specified in requirement 3. In other words, the vertical interval of contour lines needs to be changed. A rough guideline for contour intervals at different scales is summarized in Table 8.1.

In the change of contour interval, the new contour interval CI_{new} is usually a multiple of the original contour interval CI_{old}. For example, suppose the original contour interval is 2 m and the original set consists of contours at heights {2, 4, 6, 8, 10, ... 20, ...}. If the new contour interval is 10 m, then one simply select the 10s, that is {10, 20, 30, ...}. This is a simple and straightforward process. However, it is not necessarily always the case. For example, if the new contour interval is 5 m, the new contour set should consists of contours at {5, 10, 15, 20, ... In this case it is necessary to derive contours at heights of {5, 15, ...}.

TABLE 8.1
Contour Intervals at Different Map Scales

Scale	Contour Interval
1:200,000	25 to 100 m
1:100,000	10 to 40 m
1:50,000	10 to 20 m
1:25,000	5 to 20 m
1:10,000	1 to 10 m

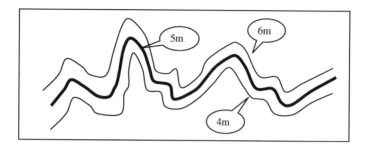

FIGURE 8.2 A new contour to be derived from two original neighboring contours (the middle line is the new one derived from the other two).

The determination of whether there is need to derive new contours can be guided by the following equation:

$$(CI_{new} // CI_{old}) \times CI_{old} \begin{cases} = CI_{new} \implies \text{no need of derivation} \\ \neq CI_{new} \implies \text{selection} + \text{derivation} \end{cases} \tag{8.1}$$

where // is an operation called integer division, that is, taking the integer part only. In the example given above, $CI_{New} = 5$ and $CI_{old} = 2$, $(5 // 2) \times 2 = 2 \times 2 = 4 \neq 5$. This indicates a need to derive some new contours. For example, the contour at 5 m may be derived from the ones at 4 m and 6 m, as shown in Figure 8.2.

One way to derive a new contour line from the original two lines is to derive the skeleton of the areas formed by the two original lines. Details on skeletonization have been discussed in Chapter 2. Experience gained from experimental testing shows that many small (unwanted) branches may be produced by skeleton algorithms (also see Section 2.4.3). The idea behind this algorithm is:

1. To construct a triangular network using the points on the two contour lines.
2. To linearly interpolate points for the new line, with the height of new contour given.
3. To join these points to form a new contour line.

In this alternative procedure, the Delaunay triangulation algorithm can be employed (see Section 2.3.1). One special consideration is the *flat triangles* that create an artificial flat area. Figure 8.3 illustrates this problem, which is common when there are spike-like lines. To avoid this problem, one possible solution is to add some points along the ridgelines or ravines into the contour data set. The other solution is to set a condition: no more than two points should be selected from a single contour line to form a triangle. More detailed discussion for solving the flat triangle problem can be found in the book by Li et al. (2005).

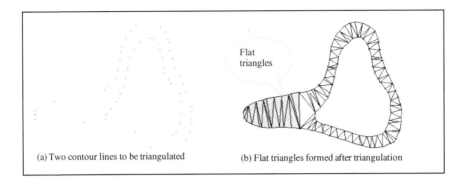

(a) Two contour lines to be triangulated   (b) Flat triangles formed after triangulation

FIGURE 8.3 A flat-triangle after a simple triangulation process (Reprinted from Li and Sui, 2000. With permission.).

8.2.3 Objective Generalization of a Set of Contour Lines as a Whole

Considering the three requirements for contour generalization described in the previous subsection, one can see that those point-reduction algorithms would not work, or, more precisely speaking, they may work only with a minimal scale change. It can also be noticed that the smoothing algorithms are not directly related to the map scale yet.

Among the available algorithms, Weibel (1996) has made a critical evaluation and found that the Li–Openshaw algorithm discussed in Chapter 7 guarantees no self-intersection. Indeed, Li and Sui (2000) have demonstrated via experimental testing that the other two requirements can also be met by this algorithm if it is carefully implemented.

For a set of contour lines, a common raster template is needed to make sure their relative positions between contour lines can be maintained. That is, it would not work well if one puts the first cell centered at the starting point of each line. Therefore, the simple implementation of the Li–Openshaw algorithm for the generalization of contour lines is to first lay down a raster template for the whole area and then to generalize each contour line with respect to the common raster template. The cell size of the raster grid is the smallest visible size (SVS). The mechanism to take a point as the representative for the spatial variations within a cell is the same as that used in the vector–raster algorithm (see Chapter 7).

In the generalization of a contour map, a special consideration needs to be taken: the size of the smallest loops (closed contour) to be retained. It is quite obvious that if the looped line is within a cell (smallest visible object [SVO]), it should be deleted. However, depending on the position of the starting point of the grid, the same line may appear on four neighboring cells.

To illustrate the performance of this procedure for contour generalization, a set of contour lines digitized from a topographic map (Li and Sui, 2000) is used. This set of contour lines depicts the terrain surface of a small island. Figure 8.4 shows

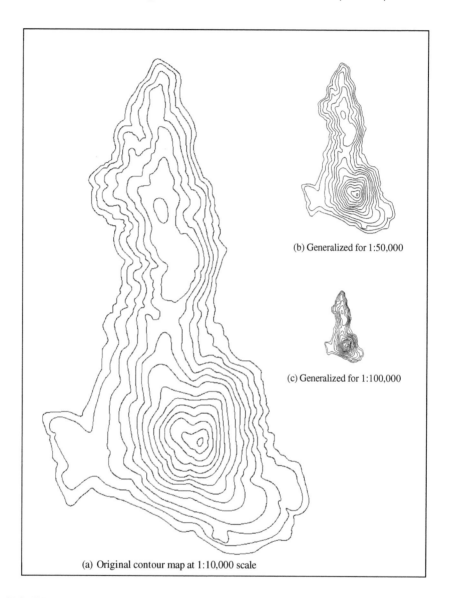

(b) Generalized for 1:50,000

(c) Generalized for 1:100,000

(a) Original contour map at 1:10,000 scale

FIGURE 8.4 Generalization of contour maps to various scales using the method developed (Reprinted from Li and Sui, 2000. With permission.).

the contour digitized from the original map at 1:10,000. The set of contour lines is generalized to suit the representations at three different scales. Figure 8.4 shows the original contour map and the generalized results at 1:50,000 and 1:100,000. The result for 1:200,000 is shown in Figure 8.5. To compare the results with the original contour map, those at 1:100,000 and 1:200,000 are then enlarged and superimposed onto the original contour map. The results are shown in Figure 8.6 and Figure 8.7.

FIGURE 8.5 Superimposition of the result generalized for a 1:100,000 scale onto the original (1:10,000) (Reprinted from Li and Sui, 2000. With permission.).

It can be seen clearly that, even for a 20 times reduction to 1:200,000, the contour lines appear to be parallel and follow the main trend very well, but small details are removed. These results depict the terrain surface nicely.

This is an evaluation in an absolute sense. Li and Sui (2000) have also conducted an experimental test in a relative sense. That is, the results from both the Li–Openshaw algorithm and another popular point-reduction algorithm were compared. This relative evaluation is omitted here because it is clear that no point-reduction algorithms are able to produce satisfactory results.

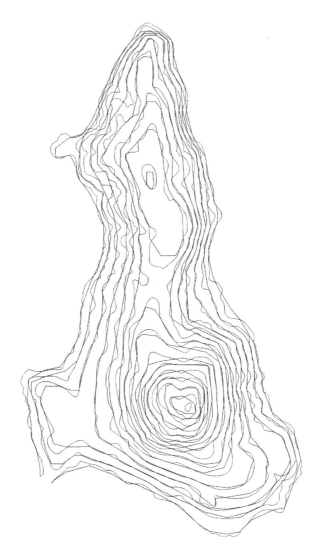

FIGURE 8.6 Superimposition of the result generalized for a 1:200,000 scale onto the original (1: 10,000) (Reprinted from Li and Sui, 2000. With permission.).

8.2.4 Transformation of a Set of Contour Lines via the Removal of Small Catchments

Ai (2004) has presented an algorithm for the transformation of contour lines as a group based on structural analysis, as follows:

1. Extract the drainage system from the contour lines.
2. Build the associations between the valley branches and the corresponding bends group of contour lines.

3. Analyze the properties of the drainage tree structure to decide which branch to remove.
4. Perform the geometric elimination of bends.

The drainage system is extracted directly from the contour lines (e.g., Yoeli, 1984; Tang, 1992) instead of from a DTM (e.g., Mark, 1984; O'Callaghan and Mark, 1984; Band, 1986; Zhu et al., 2003) in order to maintain high fidelity. This is a well-researched topic. The critical step is to decide which valley is less important so it can be removed, which is a complex process (Weibel, 1992). Many factors should be considered such as the depth, width, and area size of the valley; the order level in the drainage tree; distribution density of the valley; and separation between neighboring channels. For simplicity, Ai (2004) demonstrated the use of the following two parameters as criteria for the elimination of less important bends:

1. The valley depth.
2. The level of the valley at the catchment tree.

The process of catchment removal for contour generalization is illustrated in Figure 8.5. As Ai (2004) admitted, the graphics do not look nice. Indeed, the removal of a small catchment in such a way is overdone, although the idea behind the so-called structure-based generalization sounds good.

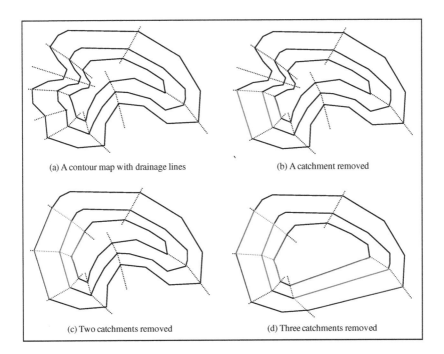

(a) A contour map with drainage lines

(b) A catchment removed

(c) Two catchments removed

(d) Three catchments removed

FIGURE 8.7 Contour generalization through the removal of small catchments.

8.3 ALGORITHMS FOR TRANSFORMATION OF RIVER NETWORKS

8.3.1 OVERVIEW

To transform the representation of a river network from a larger scale to a smaller scale, a procedure (or algorithm) consisting of the following three steps can be applied:

1. Assign an order number to each tributary to form a tree structure.
2. Omit a certain percentage of tributaries below a certain order.
3. Perform essential geometric transformations for the selected river tributaries.

Various ordering schemes for such structures have already been developed by geomorphologists and used to determine the selective omission of tributaries. This chapter is devoted to the use of such ordering systems for the selection of river networks. Four selection schemes will also be presented, but the decision on the use of a particular scheme needs to be made with the support of thematic knowledge. The discussion of such thematic knowledge lies outside the scope of this text; readers who are interested in this topic may find useful information in the book by Buttenfield and McMaster (1991).

After the selective omission operation, the selected tributaries may need to be simplified. A discussion of the transformations applied to the selected river tributaries will also be briefly discussed in this section.

8.3.2 ORDERING SCHEMES FOR SELECTIVE OMISSION OF RIVERS

In geomorphology many ordering schemes have been developed. The *Horton scheme* (Horton, 1945) and the *Shreve scheme* (Shreve, 1966) have been considered the most relevant schemes for the multi-scale representation of river networks (Rusak Mazur and Caster, 1990).

The Horton scheme assigns order 1 to all branchless fingertip tributaries and order 2 to those receiving branches or tributaries of order 1. A stream with order 3 receives one or more order 2 tributaries and possibly also some order 1 tributaries. That is, a stream with a higher order receives streams of one order lower and possibly of some even lower order. The highest order is assigned to the main trunk. This order scheme is illustrated in Figure 8.8a. This is called a downstream routine.

An upstream routine is then applied to determine which is the parent stream and to aggregate the parent stream segments into a river. This process uses two rules (Rusak Mazur and Caster, 2000):

1. Starting below the junction, extend the parent stream upstream from the bifurcation in the same direction. The stream joining the parent stream at the greatest angle is of the lower order (with the exception of some geological conditions).
2. If both streams are at about the same angle to the parent stream at the junction, the shorter one is usually assumed to be the lower order.

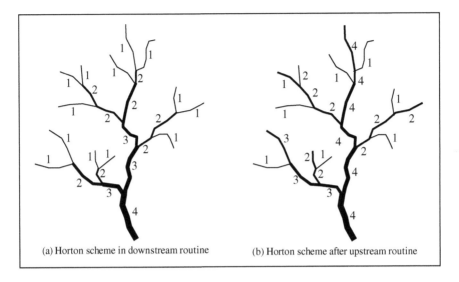

(a) Horton scheme in downstream routine   (b) Horton scheme after upstream routine

FIGURE 8.8 The Horton order scheme for a river network.

The result of such an upstream process is shown in Figure 8.8b. Rusak Mazur and Caster (2000) commented that the Horton scheme enumerates stream order, while the Shreve scheme enumerates stream magnitude. The Shreve scheme is illustrated in Figure 8.9. That is, the order of a lower stream is the sum of the orders of the upper streams.

A comparison of these two schemes made by Rusak Mazur and Caster (2000) is as follows: If the only consideration in choosing the order method is the technical criteria, Shreve's scheme would be satisfactory. Geographical and cartographic

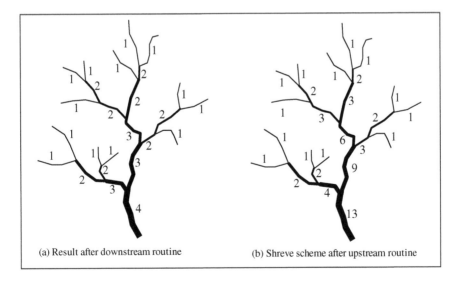

(a) Result after downstream routine   (b) Shreve scheme after upstream routine

FIGURE 8.9 The Shreve order scheme for a river network.

requirements are fulfilled by only Horton scheme because it allows the preservation of aggregated river entities or linkages under one proper name. However, the Horton scheme falls short of Shreve's scheme on technical background.

In the Horton scheme the initial classification needs to be conducted by a person knowledgeable in the area. The Shreve scheme is more straightforward, and its automation is easy to achieve. A tributary with one end node not connected is assigned an order of 1. Two or more order 1 tributaries form a stream segment with a higher order. The order of a lower stream is the sum of the orders of the upper streams. If the data are in raster format, then the hit-miss operator (see Chapter 2) can be used to detect the intersections. After that, individual stream segments can then be sorted out.

8.3.3 FOUR STRATEGIES FOR SELECTIVE OMISSION OF ORDERED RIVER FEATURES

There are four possible options for the elimination of river tributaries, as shown in Figure 8.10. In this diagram, the X-axis represents the order of the streams and the Y-axis represents the total number of streams at each order. The shadowed area indicates the portion to be eliminated.

In case A a portion of lower-order streams is included and the top order streams are all retained. In case B a much smaller portion of the very-low-order streams are

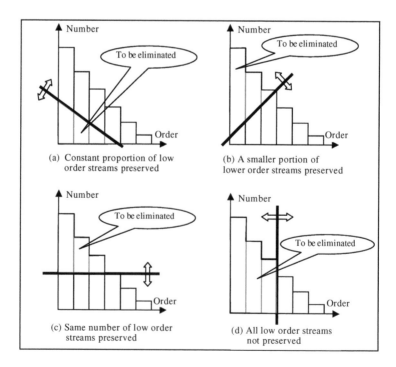

FIGURE 8.10 Four possible options for the elimination of river tributaries (Modified from Rusak Mazur and Caster, 1990.).

included and the top-order streams are all retained. In case C the same portion of low-order streams is included and all high-order streams are retained. In case D all low-order streams are eliminated.

It is clear that case D is the simplest approach. In this case a full order of low streams is eliminated so that one does not need to worry about which ones to eliminate. In the selective omission of tributaries, the principle of selection may be employed to determine the number of tributaries to be retained, and thematic rules should be formed to determine which scheme to select for a particular area.

8.3.4 OTHER TRANSFORMATIONS FOR SELECTED RIVER FEATURES

After the selective omission operation, several operations may be applied to the selected tributaries to simplify the complexity of line features.

If the rivers are already represented by a single line, the Li–Openshaw algorithm (see Section 7.4) can be applied to each tributary to simplify the structure of individual lines, and the Douglas–Peucker algorithm (see Section 5.3.2) can be applied to remove redundant points. If desirable, some algorithms for the typification of line bends may also be applied. The algorithm developed by Plazanet et al. (1995) is based on Fourier transformation, and the one by Burghardt (2005) is based the snakes technique.

If the rivers are represented by double lines at the source scale and need to be represented by a single line at the target scale, then a *collapse* operation may need to be used to transform the area represented by the double lines into a single line. McMaster and Shea (1992) consider the fusion of the two banks of a river *merging*. Nickerson (1988) has described an algorithm for such a purpose, which is the same as the medial axis transformation (see Section 2.4.1). This is a topic to be addressed in Chapter 9.

8.4 ALGORITHMS FOR TRANSFORMATION OF TRANSPORTATION NETWORKS

8.4.1 AN OVERVIEW

To transform the representation of a transportation network from a larger scale to a smaller scale, similar to transforming a river network, a procedure (or algorithm) consisting of the following three steps can be applied:

1. Assign an order number to each road segment.
2. Omit a certain percentage of road segments below a certain order.
3. Perform essential geometric transformations for the selected segments.

The selective omission of road segments is usually based on the type of road. In this way, roads may be classified into expressways, main roads, and secondary roads. Such a classification may vary from country to country and may also depend on the map scale. Figure 8.11 shows the classification scheme for topographic maps of Hong Kong from 1:1,000 to 1:200,000. At 1:20,000, the roads are classified into very detailed subclasses, but there are only two classes left at 1:200,000. This figure also indicates the class aggregation during the transformation from 1:20,000 to 1:200,000.

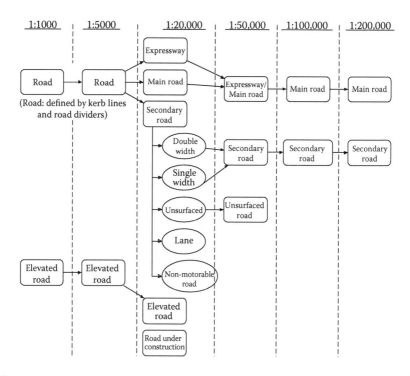

FIGURE 8.11 Transformation of road features in scale (Reprinted from Li and Choi, 2000. With permission.).

It would be an easy job to do if the whole class of road features is eliminated during a transformation process, for example, the removal of "lane" class from 1:20,000 to 1:50,000. However, this is not always the case. In most cases only a certain percentage (e.g., 40%) of features in a class is eliminated, as shown in Table 8.2. The difficulty is to determine which one to be within this percentage.

The results shown in Table 8.2 were obtained from an analysis of a portion of topographic maps of the Hong Kong Island. The values themselves might be not that representative, but they highlight the problem of selective omission in multi-scale representation of a road network.

TABLE 8.2

Change of Road Segments at Different Classes from 1:1,000 to 1:200,000 (extracted from Li and Choi, 2002)

	1:1,000	1:5,000	1:20,000	1:50,000	1:100,000	1:200,000
Main roads	100%	100%	100%	93.7%	93.7%	81.0%
Secondary roads	100%	100%	100%	55.4%	28.7%	3.6%
Total	100%	100%	100%	65.0%	46.4%	25.4%

This means that type is not sufficient for the selective omission. Each road segment needs to be assigned an order based on thematic, geometric, and topological information. Apart from type, other thematic attributes such as length, width, number of lanes, number of traffic ways, and connectivity may also be used (e.g., Li and Choi, 2000). Geometric and topological information has also been considered in the assignment of order. Mackaness and Beard (1993), Mackaness (1995), and Thomson and Richardson (1995) have used graph theory to model the topological structure of road networks. Thomson and Richardson (1999) have used the *good continuation* grouping principle to concatenate road segments into *stroke*. The stroke scheme by Thomson-Richardson will be introduced in the next subsection.

After the selective omission operation, the selected roads may need to be simplified. A discussion of the transformations applied to the selected roads will also be briefly discussed in this section.

8.4.2 THE STROKE SCHEME FOR SELECTIVE OMISSION OF ROADS

How to select a certain percentage of road segments to be retained is the key to the success of the multi-scale representation of a road network. An interesting solution is the *stroke* scheme developed by Thomson and Richardson (1999). This scheme is based on the *good continuation grouping principle* of perception. With this principle, a road network can be decomposed into a set of linear segments called strokes. These strokes are then ordered to reflect their relative importance in the network. The elimination of roads is then performed progressively according to the ordering scheme. Attributes of a road segment (arc) can be considered in the weighting related to its importance.

Figure 8.12 illustrates the principle of the stroke scheme. This figure shows 14 arcs and 13 nodes (Figure 8.12a). Based on the good continuation principle, road segments (arcs) are concatenated into chains, called strokes. In this example, the 14 arcs are concatenated into 6 strokes (Figure 8.12b).

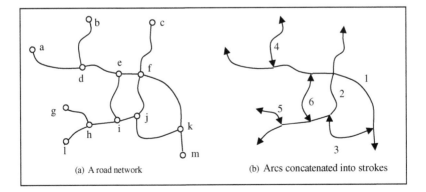

(a) A road network   (b) Arcs concatenated into strokes

FIGURE 8.12 Simple concatenation of a road network into strokes (Reprinted from Thomson and Richardson, 1999. With permission.).

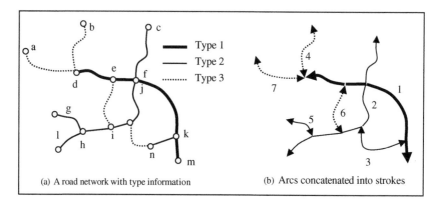

(a) A road network with type information   (b) Arcs concatenated into strokes

FIGURE 8.13 Concatenation of a road network with different types into strokes (Modified from Thomson and Richardson, 1999.).

As pointed out previously, other thematic attributes can be used as weights for the assignment of an importance value to each road segment (arc). Figure 8.13 shows the formation of strokes with consideration of road types. In this case seven strokes have been formed. In this concatenation process one principle is assumed: It is allowable to concatenate a road segment of type 3 with an arc of type 2 but not type 1.

The next step is to consider different attributes to assign a weight to each strike. The weights for different attributes based on the empirical study by Li and Choi (2002) are shown in Table 8.3, which is based on their study of Hong Kong maps. In a different country, the weighting scheme could be different.

A table such as Table 8.4 can be built for the assignment of an importance ranking to each stroke. In this example no information about number of lanes, number of traffic directions, and width is available. Therefore, the decision is made based only on type, length, and connectivity. As indicated in Table 8.4, type is most important factor, so stroke 1 is assigned order 1, although it is shorter than stroke 2. Strokes 4 and 6 have same type and same length, but stroke 6 is assigned a higher ranking because it has more connectivity numbers.

After the strokes are ranked, elimination can be made progressively. The percentage of road segments of different types to be retained can be guided by Table 8.2, similar experimental data, or the principle of selection. Figure 8.14 shows the progressive elimination of road segments for representation at different scales.

TABLE 8.3
Weights for Different Road Attributes

Attribute	Type	Length	Lane No.	Traffic directions	Width	Connections
Weight	27%	20%	18%	15%	13%	7%

TABLE 8.4
Assignment of a Ranking to Each Stroke Based on Road Attributes

Stroke #	Type	Length	Lane #	Traffic #	Width	Connection #	Ranking
1	1	40	?	?	?	2,0	1
2	2	45	?	?	?	0,0	2
3	2	17	?	?	?	2,2	3
4	3	15	?	?	?	2,0	7
5	2	9	?	?	?	2,0	4
6	3	15	?	?	?	2,2	6
7	3	16	?	?	?	1,0	5

In summary, the algorithm for selective omission of road segments using the Thomson–Richardson stroke scheme can be written as follows:

1. Concatenate road segments into strokes based on the good continuation grouping principle of perception.
2. Assign a ranking to the strokes by considering geometric, thematic, and topological information.
3. Eliminate those strokes with low ranking, as required.

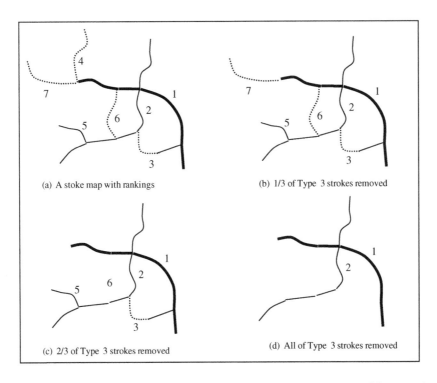

(a) A stoke map with rankings

(b) 1/3 of Type 3 strokes removed

(c) 2/3 of Type 3 strokes removed

(d) All of Type 3 strokes removed

FIGURE 8.14 Progressive elimination of road segments for representation at different scales.

8.4.3 ROAD JUNCTION COLLAPSE: RING-TO-POINT COLLAPSE

Road junction collapse is a type of *ring-to-point collapse*. Mackaness and Mackechnie (1999) presented an interesting procedure for the detection and collapse of junctions in a road network. The Mackaness–Mackechnie procedure works as follows:

1. Identify road junctions using a combination of spatial clustering and graph theory.
2. Collapse the junction by contraction with restriction of a road graph.
3. The centroid of the junction is used as its location after collapse.

A *junction* in a graph form is defined by Mackaness and Mackechnie (1999) as a relative dense collection of vertices, each with degree 3 or more. Figure 8.15 shows a simple example of such an area. In their algorithm the region with relative dense vertices is detected by clustering the x, y coordinates of the vertices, which are treated as simple points in the 2-D plane. Of the variety of clustering algorithms available, the single-linkage clustering was employed. After the detection of the junction vertices, a graph is created to reconsider their connectivity.

In the next step the position of a new vertex as the location of the junction is computed by taking the mean x, y locations (i.e., the centroid) of the connected vertices within the cluster. The attributes of each edge (road segment) may also be used for weighting in the calculation of the centroid position. Figure 8.15b shows such a new vertex. The edge involved is reattached to this new vertex.

Figure 8.15 is a very simple illustration. The practical situation of a junction could be much more complex, depending on the scale of the map. Figure 8.16 is a more complex junction, with two subnetworks separated by a water body. The graph also accordingly consists of two subgraphs. The nature of this separation will be retained after collapse. In this case the reduction of the graph will also be carried out separately. First, vertices connected solely within a subgraph are removed. Next, any repeated items are removed from these lists. In the process of aggregating vertices, the closest pair of vertices is grouped first and the position of their centroid recorded. Then, the next closest pair of vertices and centroids is grouped.

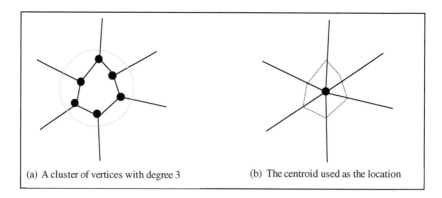

(a) A cluster of vertices with degree 3   (b) The centroid used as the location

FIGURE 8.15 The principle of the Mackaness–Mackechie procedure for junction collapse.

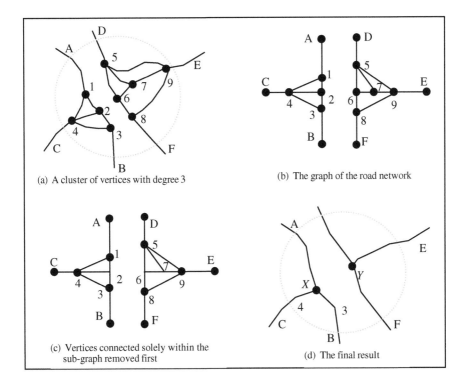

(a) A cluster of vertices with degree 3

(b) The graph of the road network

(c) Vertices connected solely within the sub-graph removed first

(d) The final result

FIGURE 8.16 Collapse of a complex junction.

8.4.4 OTHER TRANSFORMATIONS FOR SELECTED TRANSPORTATION LINES

After the selective omission operation, several operations may be applied to the selected road segments to simplify the complexity of line features. Some of these operations are identical to those for the simplification of river features.

If the roads are already represented by a single line, the Li–Openshaw algorithm (see Section 7.4) can be applied to each segment to simplify the structure of individual lines, and the Douglas–Peucker algorithm (see Section 5.3.2) can be applied to remove redundant points. If desirable, some algorithms for the typification of line bends may also be applied. The algorithm developed by Plazanet et al. (1995) is based on Fourier transformation, and the one by Burghardt (2005) is based on the snakes technique.

If the road segments are represented by double lines at the source scale and need to be represented by a single line at the target scale, then a *double-to-single-line collapse* operation may need to be used to transform the area represented by the double lines into a single line. Nickerson (1988) has described an algorithm for such a purpose, which is the same as the medial axis transformation (see Section 2.5.1). The skeletonization algorithms described in Section 2.5 of Chapter 2 and the algorithms for "area-to-line collapse" discussed in Section 9.4.1 of Chapter 9 can be used for such a purpose.

Sometimes, it is also necessary to produce a combined representation of many tracks at a railway station. McMaster and Shea (1992) considered this *merging*. The skeletonization algorithms described in Section 2.5 and the algorithms for area-to-line collapse discussed in Section 9.4.1 can also be used for this.

When the road junction is reduced for representation at a smaller scale, the characteristics (i.e., two lines not aligned) are not clear anymore and need to be enhanced (Table 1.5). This kind of operation is referred to as *enhancement* in this text although it is sometimes called *exaggeration* in other literature. Such an enhancement can be achieved by displacing any or both of these two lines. The topic of displacement will be discussed in Chapter 11.

REFERENCES

Ai, T. H., A generalization of contour line based on the extraction and analysis of drainage system, *International Archive of Photogrammetry and Remote Sensing*, XXXV(IV/3), 2004. (CR-ROM)

Band, L. E., Topographic partition of watersheds with digital elevation models, *Water Resource Research*, 22(1), 15–24, 1986.

Burghardt, D., Controlled line smoothing by snakes, *GeoInformatica*, 9(3), 237–252, 2005.

Buttenfield, B. P. and McMaster R. B., (eds.), *Map Generalization: Making Rules for Knowledge Representation*. Longman Scientific and Technical, London, 1991.

Chen, Z.-T., Contour generalization by a 3-dimensional spatial low-pass filtering, in *Proceedings of GIS/LIS'87*, San Francisco, 1987, pp. 375–386.

Horton, R. E., Erosional development of streams and their drainage basins, *Bulletin of the Geological Society of America*, 56, 275–370, 1945.

Li, Z. L. and Choi, Y. H., Topographic map generalisation: association of road elimination with thematic attributes, *Cartographic Journal*, 39(2), 153–166, 2002.

Li, Z. L. and Openshaw, S., Algorithms for automated line generalisation based on a natural principle of objective generalisation, *International Journal of Geographical Information Systems*, 6(5), 373–389, 1992.

Li, Z. L. and Sui, H. G., An integrated technique for automated generalisation of contour maps, *Cartographic Journal*, 37(1), 29–37, 2000.

Li, Z. L., Zhu, Q., and Gold, C., *Digital Terrain Modelling: Principles and Methodology*, CRC Press, Boca Raton, FL, 2005.

Mackaness, W. A., Analysis of urban road networks to support cartographic generalization, *Cartography and Geographic Information Systems*, 22(4), 306–316, 1995.

Mackaness, W. and Mackechnie, G. A., Automating the detection and simplification of junctions in road networks, *GeoInformatica*, 3(2), 185–200, 1999.

McMaster, R. B. and Shea, K. S., *Generalization in Digital Cartography*, Association of American Geographers, Washington, DC, 1992, 134 pp.

Mark, D. M., Automated detection of drainage network from digital elevation model, *Cartographica*, 21(3), 168–178, 1984.

Nickerson, B., Automated cartographic generalization for line features, *Cartographica*, 25(3), 15–66, 1988.

Nickerson, B. and Freeman, H., Development of a rule-based system for automatic map generalization, in *Proceedings of the Second International Symposium on Spatial Data Handling*, 1986, pp. 537–556.

O'Callaghan, J. F. and Mark, D. M., The extraction of drainage networks from digital elevation data, *Computer Vision, Graphics and Image Processing*, 28(3), 323–344, 1984.

Plazanet, C., Affholder, J. G., and Fritsch, E., The importance of geometric modelling in linear feature generalization, *Cartography and Geographic Information Systems*, 22(4), 291–305, 1995.

Rusak Mazur, E. and Caster, H. W., Horton ordering scheme and the generalisation of river networks, *Cartographic Journal*, 27(2), 104–112, 1990.

Shapiro, B., Pisa, J., and Sklansky, J., Skeleton generation from x,y boundary sequences, *Computer Graphics and Image Processing*, 15,136–153, 1981.

Shreve, R. L., Statistical law of stream numbers, *Journal of Geology*, 74, 17–37, 1966.

Tang, L., Automatic extraction of specific geomorphologic elements from contours, in *Proceedings of the 5th Symposium on Spatial Data Handling*, Vol. 2, Charleston, SC, 1992, pp. 554–566.

Thomson, R. C. and Richardson, D. E., A graph theory to road generalization, in *Proceedings 17th International Cartographic Conference*, 1995, pp. 1871–1880.

Thomson, R. C. and Richardson, D. E., 1999. The 'good continuation' principle of perceptual organization applied to the generalization of road network, in *Proceedings of ICC'01*, August 2001, Ottawa. CD-Rom. (also available at http://ccrs.nrcan.gc.ca/radar/map/genrn_e.php)

Wang, Z. and Muller, J. C., Line generalization based on analysis of shape, *Cartography and Geographic Information Systems*, 25(1), 3–15, 1998.

Weibel, R., An adaptive methodology for automated relief generalization, in *Proceedings of Auto-Carto 8*, Baltimore, MD, 1987, pp. 42–49.

Weibel, R., Model and experiments for adaptive computer-assisted terrain generalization, *Cartography and Geographic Information Systems*, 19(3), 133–153, 1992.

Weibel, R., A typology of constraints to line simplification, in *Proceedings of 7th International Symposium on Spatial Data Handling (SDH '96)*, 1996, pp. 9A.1–9A.14.

Weibel, R., Generalization of spatial data: principles and selected algorithms, in *Algorithmic Foundation of Geographic Information Systems*, van Kreveld, M., Nievergelt, J., Roos, T., and Widmayer, P., Eds., Springer, Berlin, 1997, pp. 99–152.

Werner, C., Formal analysis of ridge and channel pattern in maturely eroded terrain, *Annals of the Association of American Geographers*, 78(2), 253–270, 1988.

Wu, H., Structured approach to implementing automated cartographic generalisation, in *Proceedings of ICC'97*, Vol. 1, Stockholm, 1997, pp. 349–356.

Yoeli, P., Computer-assisted determination of the valley and ridge lines of digital terrain models, *International Yearbook of Cartography*, XXIV, 197–206, 1984.

Zhu Q., Zhao J., Zhong Z., and Sui H. G., An efficient algorithm for the extraction of topographic structures from large scale grid DEMs, in *Advances in Spatial Analysis and Decision Making*, Li, Z. L., Zhou, Q., and Kainz, W., Eds., A.A. Balkema, Lisse, The Netherlands, 2003, pp. 99–107.

9 Algorithms for Transformations of Individual Area Features

In the previous four chapters we presented algorithms for lines features. In this and the next chapters the discussion will focus on area features. This chapter will discuss algorithms for individual features.

9.1 TRANSFORMATION OF INDIVIDUAL AREA FEATURES: AN OVERVIEW

In Chapter 1 a list of six operations was provided for the transformation of individual area features: elimination, (shape) simplification, collapse, split, exaggeration, and displacement (see Table 1.6). Displacement is a common operation for point, line, and area features and will be discussed in Chapter 11. The transformations of the variation in the third dimension will be discussed in Chapter 12.

An area feature can be represented by a *region* (a group of pixels) in raster or its *boundary* in vector. Some algorithms are based on the manipulation of the vector boundary of the area feature, leading to the term *boundary-based shape simplification*, while the others are based on the manipulation of the region covered by the area feature, leading to the term *region-based shape simplification*. Algorithms for boundary-based shape simplification will be discussed in Section 9.2. and algorithms for region-based shape simplification will be presented in Section 9.3.

Collapse is a transformation with a change in the dimensionality of an area feature. According to dimensionality, there are three types of collapse: area-to-point, area to line, and partial. In this book, linear features (such as rivers and roads) represented by double lines are treated as thin area features. Section 9.4 will be dedicated to the discussion of this topic.

There are two types of elimination for an area feature. The first type is the elimination of a simple polygonal area, and the second is the elimination of an area within another larger area, such as a hole. The second type of elimination is sometimes called *coarsening* (Beard and Mackaness, 1991). Section 9.5 will discuss this topic.

Splitting is breaking the bottleneck connection between two or more larger parts of an area feature. This operation can be easily realized in raster mode by morphological operators. Section 9.6 will discuss this. The last topic is related to the so-called exaggeration. Three types have been identified: enlargement, enhancement, and partial thickening. This operation will be discussed in Section 9.7.

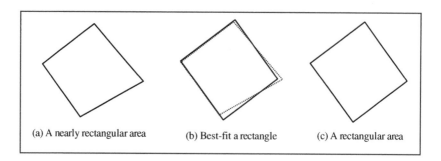

FIGURE 9.1 Rectification of an area feature.

It must be noted here that in some of the literature rectification as shown in Figure 9.1 is regarded as an operation for multi-scale representation of area features. However, the author takes a different view and restrictedly treats it as an editing operation. As a result, it is left out of this chapter.

9.2 ALGORITHMS FOR BOUNDARY-BASED SHAPE SIMPLIFICATION OF AN AREA FEATURE

9.2.1 BOUNDARY-BASED AREA SHAPE SIMPLIFICATION: NATURAL VERSUS EXTREMAL

Boundary-based area simplification simplifies the shape of an area feature through a change in the complexity of the boundary. The level of simplification depends the target scale of the representation.

If the scale change is not very large or the target scale is not very small, the representations can be more natural. The algorithms presented in Chapters 5 to 7 may be applied in such cases. If the target scale is very small, then the representation can be extremely simplified, leading to the term *extremal representations by bounding regions.*

Bounding region is an important concept for defining the shape of a feature and can be used to simplify individual area features. Bounding region is also called *bounding containers*. A simple classification of bounding regions is shown in Figure 9.2.

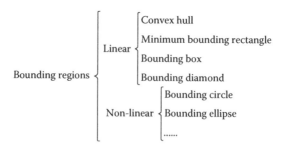

FIGURE 9.2 A classification of bounding regions.

An excellent brief description of these bounding regions has been provided by Sunday (2005). Among them, the following three types of bounding regions are popular and will be discussed in this section:

- Convex hull: The convex polygon of the smallest area that encloses the given area feature.
- Bounding box: The smallest rectangle that encloses the given area feature, which is oriented with respect to the coordinate axis.
- Minimum bounding rectangle (MBR): The smallest rectangle that encloses the given area feature.

Figure 9.3 shows an example of each of these bounding regions. The convex hull of an area feature (*A*) is the convex polygon with the smallest area that encloses area feature (*A*). The internal angle formed at each vertex of the convex hull is smaller than 180°. This means that when one walks along the boundary of the *convex polygon*, one would make only right turns if clockwise and only left turns if counterclockwise. The MBR is a specific type of convex hull, with convexity of 270° for the four external angles. MBR is also sometimes called minimum area rectangle, minimum-area enclosing rectangle, smallest-area enclosing rectangle, or minimum circumference rectangle. The bounding box is a particular type of MBR that is, it is oriented with respect to the coordinate axis.

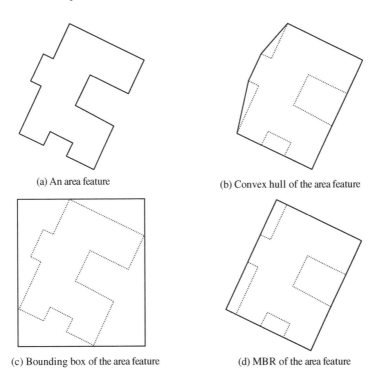

(a) An area feature   (b) Convex hull of the area feature

(c) Bounding box of the area feature   (d) MBR of the area feature

FIGURE 9.3 Bounding regions of an area feature.

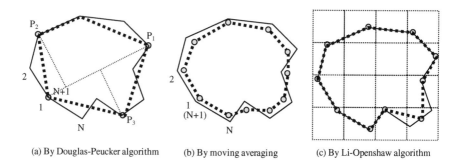

(a) By Douglas-Peucker algorithm   (b) By moving averaging   (c) By Li-Openshaw algorithm

FIGURE 9.4 Three types of boundary-based shape simplification.

In this section the applicability of the algorithms presented in Chapters 5 and 6 to boundary-based area simplification is first discussed, followed by some very simple algorithms for the generation of both convex hull and MBR.

9.2.2 NATURAL SIMPLIFICATION OF THE BOUNDARY OF AN AREA FEATURE AS A CLOSED CURVE

The boundary is simply treated as a closed curved line. The algorithms for the manipulation of line features discussed in Chapter 5 to 7 may be applied, depending on the purpose. In certain cases the $(N + 1)$th point will be identical to the first point if there are N points on the boundary (Figure 9.4).

For natural boundaries such as coastlines, administrative divisions along natural features, and boundaries of vegetation, the scale-driven generalization algorithms discussed in Chapter 7 are preferred. For those with less complexity, the algorithms presented in Chapter 6 might work. However, the parameters are not directly related to the scale of spatial representation. In some other cases such as the cadastral division of land lots, the boundaries are artificially defined and rather simple, so that the algorithms described in Chapter 5 will work well. Figure 9.4a is the result of the Douglas–Peucker algorithm; Figure 9.4b is the result of moving averaging (using the weighted averaging of three points).

It was pointed out in Section 7.4 and demonstrated in Figure 7.13 that the Li–Openshaw algorithm can be implemented for simplifying artificial lines. It must be emphasized here that the Li–Openshaw algorithm can also be implemented for simplifying artificially defined boundaries, by using the implementation demonstrated in Figure 7.13. That is, each point on the boundary is considered and is mapped to a cell. If there is only one point in a cell, this point is retained; otherwise, the averages of the points in the cell (or any among them) are taken as the representative point in the cell. Figure 9.4c shows such an implementation. It can be seen that, in this figure, only the cell at the lower-right corner contains two points, and their average is taken as the representative.

9.2.3 FORMATION OF THE CONVEX HULL OF AN AREA FEATURE

Many algorithms are available for the generation of a convex hull for a set of points, and some popular ones will be presented in Section 10.2 for the combination of area features. Usually such algorithms include a step of data sorting and another step of convexity verification. For an area feature, the points on the boundary are in order already, so there is no need to carry out a data sorting process. That is, only the convexity verification step is applicable when forming the convex hull of an area feature.

One algorithm, called the *three-coins algorithm*, developed by Graham (1972), seems to be very appropriate for the generation of a convex hull for an area feature. The algorithm, simplified for this purpose, is as follows:

1. Make sure the points on the boundary are in clockwise order.
2. Place three coins on vertices 1, 2, and 3 in clockwise order and label them "back," "center," and "front," respectively.
3. If the three coins form a right turn (or lie on a straight line), move the set of three coins forward by one point.
4. If the three coins form a left turn, remove the point at center coin from the list and push both the 'center' and 'back' coins backward by one point.
5. Repeat step 3 and 4 until the front coin is on the first point, that is, point 1, and the three coins form a right turn.
6. Join the retained points in order to form a convex hull.

Figure 9.5 illustrates the principle of the Graham scan algorithm. It can be seen clearly that the set of three coins advances along the ordered vertices as long as they

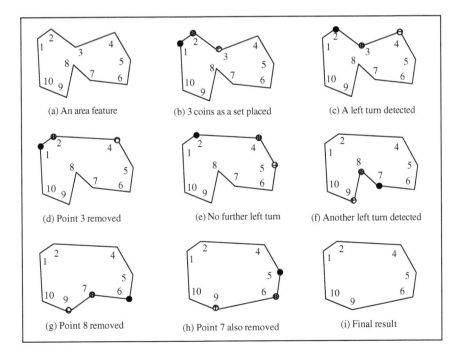

(a) An area feature

(b) 3 coins as a set placed

(c) A left turn detected

(d) Point 3 removed

(e) No further left turn

(f) Another left turn detected

(g) Point 8 removed

(h) Point 7 also removed

(i) Final result

FIGURE 9.5 Graham scan algorithm for the generation of a convex hull for an area feature.

keep forming right turns (Figures 9.5b and 9.5e). However, a left turn is encountered when the set is centered at vertex 3 (Figure 9.5c); the "guilty" point, that is, vertex 3, is then removed and both the center and back coins are pushed backward by one point to vertices 2 and 1, respectively (Figure 9.5d). The set of coins at the three new positions (i.e., vertices 1, 2, and 4) forms a right turn and the process continues (Figure 9.5e). When the set is centered at vertex 8, a left turn is formed (Figure 9.5f), and then point 8 is removed and both the center and back coins are moved backward to vertices 7 and 6, respectively (Figure 9.5g). Again, a left turn is formed, and then point 7 is also removed, and the center and back coins are moved backward to vertices 6 and 5, respectively (Figure 9.5h). After that, no more left turns are found, and the convex hull is formed (Figure 9.5i).

9.2.4 FORMATION OF THE MBR OF AN AREA FEATURE

A *bounding box*, as shown in Figure 9.3, is very straightforward to compute, that is, to sort out the four extrema among the coordinates of boundary points: X_{min}, X_{max}, Y_{min}, and Y_{max}.

Many algorithms are available for the computation of an MBR for an area feature (de Berg et al., 2000). In this section the well-known *rotating calipers* algorithm developed by Toussaint (1983) will be described. This algorithm was designed to generate the MBR of an area from its convex hull, which was discussed in Section 9.2.2. That is, the three-coins algorithm should be applied to the area boundary first if the area is not convex.

In order to make sure the bounding rectangle has a minimum area, a set of such rectangles should be examined in sequence. To make the algorithm work efficiently, the following crucial theorem has been used, which has been approved by Freeman and Shapira (1975): *The minimum bounding rectangle of a convex polygon has a side collinear with one of the edges of the polygon.*

The basic idea of Toussaint rotating callipers is (a) to start with a bounding rectangle and (b) to rotate it along the edge of the area sequentially. The issue is where to start and how to rotate efficiently. The algorithm works as follows:

1. Find the four extrema among the coordinates of boundary points: X_{min}, X_{max}, Y_{min}, and Y_{max} to form a bounding box.
2. Compute the four angles, each formed by a rectangle side and the next polygon edge in clockwise order.
3. Select the edge with the smallest angle computed in step 2 as the initial orientation of the rectangle.
4. Orientate the rectangle along the next polygon edge.
5. Enlarge or reduce the size of the rectangle to make sure it just bounds the polygon.
6. Compute the area.
7. Repeat steps 4 to 6 until the entire polygon is scanned.

The four parallel lines of rectangle edges form two pairs of calipers. This is why this algorithm is called *rotating calipers*. Figure 9.6 shows the process. Figure 9.6b

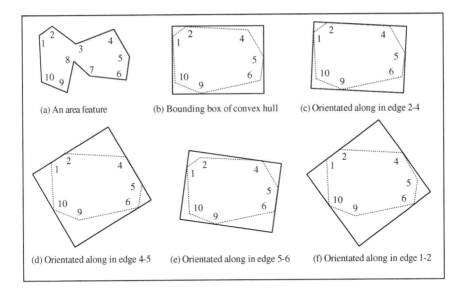

(a) An area feature   (b) Bounding box of convex hull   (c) Orientated along in edge 2-4

(d) Orientated along in edge 4-5   (e) Orientated along in edge 5-6   (f) Orientated along in edge 1-2

FIGURE 9.6 Toussaint rotating callipers for generating the MBR of an area feature.

shows the bounding box. The angle formed by the upper edge of the rectangle and polygon edge $2 - 4$ is the smallest, and thus, the first rectangle is oriented along edge $2 - 4$. The next step is to rotate the rectangle along the edge $4 - 5$, but with proper fitting of the rectangle. The process continues until all edges are scanned. It is clear that the area size varies with the orientation. In the end, the one with the smallest area size is selected as the MBR. An excellent animation of this algorithm can be found on Hormoz Pirzadeh's homepage at http://cgm.cs.mcgill.ca/~orm/welcome.html.

It is easy to say that the rectangle is rotated and enlarged or reduced to fit into the convex hull, but it is difficult to implement. In the author's view, a simpler solution is to rotate the coordinate system instead of rotating the rectangle. If the selected next polygon edge is used as one of the new coordinate axes, then the bounding rectangle is formed by the new X_{min}, X_{max}, Y_{min}, and Y_{max} after a rotation, as shown in Figure 9.7. Therefore, steps 4 and 5 could be rewritten as follows:

4. Rotate the coordinates system using the selected next edge as new x-axis and transform all vertices into the new coordinate system.
5. Identify the new X_{min}, X_{max}, Y_{min}, and Y_{max} to form a bounding rectangle.

9.3 ALGORITHMS FOR REGION-BASED SHAPE SIMPLIFICATION OF AN AREA FEATURE

The previous section introduced some boundary-based algorithms for shape simplification of an area feature. This section will describe some region-based algorithms.

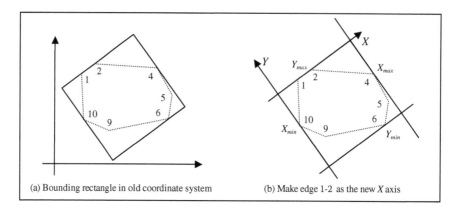

(a) Bounding rectangle in old coordinate system   (b) Make edge 1-2 as the new X axis

FIGURE 9.7 The new X_{min}, X_{max}, Y_{min}, and Y_{max} (after rotation) form the bounding rectangle.

9.3.1 SHAPE SIMPLIFICATION BY MORPHOLOGICAL CLOSING AND OPENING

Li (1994) demonstrated that a simple solution for region-based simplification of an area feature is the closing operation operator in mathematical morphology. That is, the area feature is first dilated by a selected structuring element and then eroded by the same structuring element. Figure 9.8 shows an example of using a closing operator for boundary smoothing. In this figure, the symbol + means the pixel is a result of dilation.

Not only can the closing operator be used to produce the simplification effect, but different combinations of dilation and erosion are possible for the simplification of the area boundary, although the result would be slightly different. Figure 9.9 shows an example of applying opening and closing operators in different sequences. The results are slightly different. A higher degree of simplification effect has been produced by the combination of closing and then opening operators, as shown in Figure 9.9c.

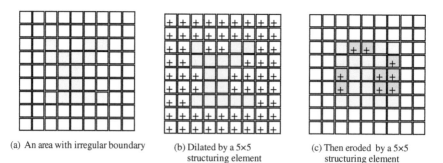

(a) An area with irregular boundary   (b) Dilated by a 5×5 structuring element   (c) Then eroded by a 5×5 structuring element

FIGURE 9.8 Closing operator for boundary smoothing.

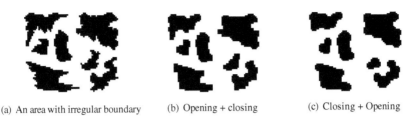

(a) An area with irregular boundary   (b) Opening + closing   (c) Closing + Opening

FIGURE 9.9 Different combinations of opening and closing for boundary simplification (Reprinted from Su et al., 1997a. With permission.).

9.3.2 FORMATION OF CONVEX HULL AND BOUNDING BOX BY MORPHOLOGICAL THICKENING

In the case of Figure 9.8, a very satisfactory result is produced by a simple closing operation. However, this rarely happens in practice. Figure 9.10a is a typical example of a settlement to be simplified, and such a simple closing operation would not work. A possible solution is to produce a bounding box as shown in Figure 9.10b. The mathematical express for the generation of such a bounding box using morphological operators is as follows:

$$BB = A \odot \{B_i\} \tag{9.1}$$

where $\odot$ is an operator called *thickening*, through which the original area image will become thicker and $\{B_i\}$ is a series of special structuring elements for thickening purposes, which are shown in Figure 9.11. The origin of each structuring element must be a white (or zero) pixel. This formula means that the bounding box is formed through sequential thickening by a set of structuring elements $\{B_i\}$.

The thickening operator is a combination of a number of other morphological operators. It is defined as follows (Serra, 1982):

$$T_k = A \odot B = A \cup (A \otimes B) \tag{9.2}$$

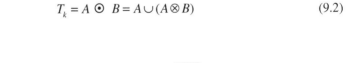

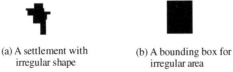

(a) A settlement with irregular shape   (b) A bounding box for irregular area

FIGURE 9.10 Simplification of a settlement with irregularity using a bounding box (Reprinted from Su et al., 1997b).

FIGURE 9.11 The series of structuring elements $\{B_i\}$ for bounding box formation (x means "don't care").

where, T_k is the result of thickening, $\cup$ is the union set operator, and $\otimes$ is the hit_miss operator.

From Equation 9.2 it can be seen that the thickened image consists of the original image plus any additional foreground pixels switched on by the hit-and-miss transform. In reality, the thickening process is also quite similar to that of dilation or erosion and is illustrated in Figure 9.12. This is achieved by moving the template of the structuring element along each pixel to find the exact matches of patterns. If a match is found, then the pixel at the position of the origin of the structuring element is changed from white to black. It must be emphasized here that the origin of the structuring element used in thickening must be white (or zero) to make the thickening possible.

In practice, a more useful expression of thickening is the so-called *homotopic sequential thickening,* which makes use of a sequence of homotopic structuring elements $\{C_i\}$ as shown in Figure 9.13. A convex hull will be generated by homotopic sequential thickening. Figure 9.14 shows this case. The numbers in Figure 9.14c indicate the structuring elements in the sequence that contributes to the thickening of this pixel.

9.3.3 SHAPE REFINEMENT BY MORPHOLOGICAL OPERATORS

The use of a bounding box for settlement representation is usually oversimplified, and the appearance is very distorted. Such a solution is only suitable for shapes that are very close to rectangular or where a rectangular shape is required for the final result (e.g., in the case of a block of buildings). It does not necessarily work well for other cases. Therefore, more robust algorithms need to be developed. For such a purpose, Su et al. (1997b) have developed an

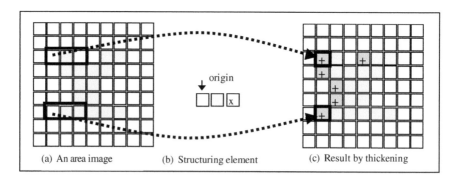

FIGURE 9.12 Thickening operation.

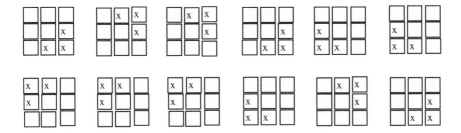

FIGURE 9.13 The series of structuring elements $\{C_i\}$ for homotopic sequential thickening (x means "don't care").

algorithm (called the *SLLM shape refinement algorithm* here for Su, Li, Lodwick, and Müller) as follows:

1. Eliminate the small convex areas on feature C by using an opening operator:

$$D = C \circ B_s \tag{9.3}$$

2. Form the bounding box of the opened area D:

$$BB_D = D \odot \{B_i\} \tag{9.4}$$

3. Obtain the complementary set of the opened area within the convex area:

$$E = BB_D - D \tag{9.5}$$

4. Eliminate small convex areas on the area E by using an opening operator:

$$F = E \circ B \tag{9.6}$$

5. Create the bounding box of the opened area F:

$$BB_f = F \odot \{B_i\} \tag{9.7}$$

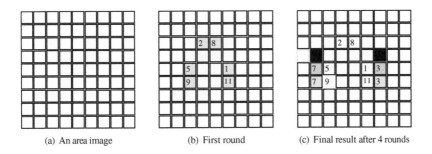

(a) An area image   (b) First round   (c) Final result after 4 rounds

FIGURE 9.14 Convex hull formation by sequential thickening.

(a) A settlement with   (b) Simplified by SLLM algorithm
    irregular shape

FIGURE 9.15 Shape refinement by the SLLM algorithm (Reprinted from Su et al., 1997.).

6. Obtain the complementary set of the opened area with the bounding box:

$$G = BB_f - F$$

7. Eliminate small convex areas on the area G by using an opening operator:

$$H = G \circ B_s \qquad (9.8)$$

8. Obtain the complementary set of H:

$$I = BB_f - H \qquad (9.9)$$

9. Obtain the complementary set of I:

$$J = BB_D - I \qquad (9.10)$$

where B_s is a structuring element that is different from $\{B_i\}$. Here, the size of B_s should be dependent on the scales of the source map and the target map. Figure 9.15 shows the results obtained with this algorithm.

9.4 ALGORITHMS FOR COLLAPSE OF AREA FEATURES

This section will present a set of algorithms for the transformation of an individual area feature with a change in dimensionality. This operation is referred to as *collapse*. Collapse is an operation used to transform area features to point features, or to transform area features completely or partially to linear features, to suit the representation at a smaller scale. Accordingly, a collapse operation can be categorized into one of two types: complete collapse and partial collapse. Complete collapse is composed of another two types: area-to-point collapse and area-to-line collapse. This section will discuss algorithms for these three types of collapse.

9.4.1 AREA-TO-POINT COLLAPSE

The *area-to-point collapse* operation transforms an area feature into a point feature when it appears to be too small to be represented as an area feature at a smaller

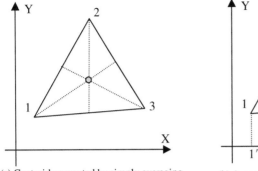
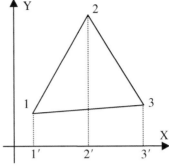

(a) Centroid computed by simply averaging   (b) Area computed from three trapziods

FIGURE 9.16 The centroid and area of a triangle.

scale. Normally, the resultant point features are positioned at the center of gravity of the original area feature.

The center of gravity is also called the *centroid* or *center of mass*. The gravity center of a triangle is simply the average coordinates of the three vertices (Figure 9.16):

$$X_{C,3} = \frac{X_1 + X_2 + X_3}{3}$$
$$Y_{C,3} = \frac{Y_1 + Y_2 + Y_3}{3}$$
(9.11)

For an ordinary area feature (polygon) with N vertices, the centroid can be computed as the weighted average of the centroids of the $(N - 2)$ triangles, which form a tessellation of the polygon. Therefore, a simple algorithm could be written as follows:

1. Select any vertex as the common vertex for all the $(N - 2)$ triangles (Figure 9.17a).
2. Compute the coordinates $(X_{C,i}, Y_{C,i})$ of the centroid for each of the triangles (i.e., $i = 0, N - 2)$) using Equation 9.11.
3. Compute the area of each triangle, that is, $A_{\Delta,i}$, using Equation 9.13 or 9.17, below.
4. Calculate the area of the whole polygon, that is, $A_{Whole} = \Sigma_{i=1}^{N-1} A_{\Delta,i}$.
5. Compute the weight for each of the triangles, that is, $w_i = A_{\Delta,i}/A_{Whole}$.
6. Compute the weighted averages as the centroid, using Equation 9.12.

$$X_C = \sum_{i=1}^{N-1} w_i \times X_{C,i}$$
$$Y_C = \sum_{i=1}^{N-1} w_i \times Y_{C,i}$$
(9.12)

The area of a triangle A_Δ can be computed by Heron's formula as follows:

$$A_\Delta = \sqrt{s(s-a)(s-b)(s-c)} \tag{9.13}$$

where a, b, and c are the three triangle sides, and s is the half-perimeter and is computed as follows:

$$s = \frac{1}{2}(a+b+c) \tag{9.14}$$

A more convenient method can be used for the computation of the area of a triangle with three given points. Figure 9.16b shows the principle. In this figure the three vertices are points 1, 2, and 3. If these three points are projected to the x-axis, then points 1', 2', and 3' are obtained. Points 1 and 2 together with 1' and 2' form a trapezoid. Points 2 and 3 together with 2' and 3' form another trapezoid, and points 3 and 1 together with 3' and 1' form the third trapezoid. By adding the areas of the first two trapezoids together and substracting the area of third trapzoid, the area of the triangle $\Delta 123$ is obtained:

$$A_{\Delta 123} = \left| A_{122'1'} \right| + \left| A_{233'2'} \right| - \left| A_{311'3'} \right| \tag{9.15}$$

However, if the vertices are arranged clockwise and the areas are computed according to Equation 9.6, the value of $A_{311'3'}$ will be negative, and then Equation 9.15 could be written as:

$$A_{122'1'} = \frac{y_1 + y_2}{2} \times (x_2 - x_1)$$

$$A_{233'2'} = \frac{y_2 + y_3}{2} \times (x_3 - x_2) \tag{9.16}$$

$$A_{311'3'} = \frac{y_3 + y_1}{2} \times (x_1 - x_3)$$

$$
\begin{aligned}
A_{\Delta 123} &= A_{122'1'} + A_{233'2'} + A_{311'3'} \\
&= \frac{1}{2}\left[(y_1 + y_2)(x_2 - x_1) + (y_2 + y_3)(x_3 - x_2) + (y_3 + y_1)(x_1 - x_3) \right] \\
&= \frac{1}{2}(y_1 x_2 + y_2 x_3 + y_3 x_1 - x_1 y_2 - x_2 y_3 - x_3 y_1) \\
&= \frac{1}{2}\begin{vmatrix} x_1 & y_1 & 1 \\ x_2 & y_2 & 1 \\ x_3 & y_3 & 1 \end{vmatrix}
\end{aligned}
\tag{9.17}
$$

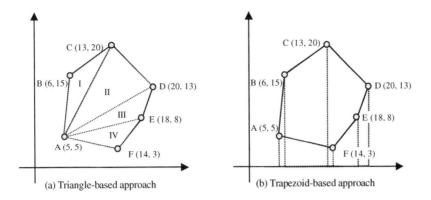

(a) Triangle-based approach   (b) Trapezoid-based approach

FIGURE 9.17 Two ways to compute the area of an area feature (polygon).

Figure 9.17a shows an example of using this approach to compute the centroid of an area feature. In this example, there are six points, whose coordinates are shown. Point A is selected as the common point. From this point to other vertices, four (= 6 – 2) triangles are formed. The centroids for these four triangles are (8, 13.3), (12.7, 12.7), (14.3, 8.7), and (12.3, 5.3). The areas are 32, 80, 29, and 27, respectively. The total area is 168. The weights are thus 0.19, 0.48, 0.17, and 0.16, respectively. Therefore, the centroid of this area feature is (12, 11).

Equation 9.17 can be extended to compute the area of any polygon with N points, as follows:

$$A = \frac{1}{2} \sum_{i=1}^{N} (y_i \times x_{i+1} - x_i \times y_{i+1}) \tag{9.18}$$

This formula requires the $(N + 1)$th point. However, it does not exit in the point list of the polygon. As a result, the first point is used as the $(N + 1)$th point so as to make this polygon closed. Using the same approach, the centroid of a simple polygon can be computed as follows (Bourke, 1988):

$$X_C = \frac{1}{6A} \sum_{i=1}^{N} (y_i \times x_{i+1} - x_i \times y_{i+1})(x_i + x_{i+1})$$

$$Y_C = \frac{1}{6A} \sum_{i=1}^{N} (y_i \times x_{i+1} - x_i \times y_{i+1})(y_i + y_{i+1}) \tag{9.19}$$

The centroid of the area feature shown in Figure 9.17 computed by using Equation 9.19 is identical to that obtained by using Equation 9.12.

9.4.2 AREA-TO-LINE COLLAPSE*

The area-to-line collapse happens when an area is a long but thin feature. In this case the feature will appear to be too thin to be represented as an area feature at a

---

* This section is largely extracted from Su et al., 1998. With permission.

smaller scale. As a result, such features will be represented as line features at a smaller scale. The main task for this operation is to derive an appropriate line to represent the original area feature. Among many potential candidates, the skeleton of the original area feature seems to be a logical choice. This candidate has been widely accepted, and derivation of appropriate skeletons has been considered by several researchers (DeLucia and Black, 1987; Chithambaram et al., 1991; Jones et al., 1995).

Skeletonization was addressed in Section 2.4. Three approaches were presented:

1. By means of medial axis transformation and distance transformation.
2. By means of Voronoi diagram and triangulation.
3. By means of morphological thinning.

As discussed in Section 2.4, by applying the homotopic sequential thinning and pruning algorithms in a sequence, the skeletons of area features can be derived. However, there is no solution for automated control over loop number N in the pruning process expressed by Equation (2.4). Through a number of experiments, Su et al. (1998) found that it is possible to overcome this deficiency by using a subset of $\{P_i\}$ shown in Figure 2.8 and by integrating the thinning and pruning processes. More exactly, they suggest the use of only the last four structuring elements in $\{P_i\}$. As a result, the complete skeletonization algorithm for the area-to-line collapse is as follows:

$$SK(A) = A \bigcirc \{B_i\} \bigcirc \{E_i\} \tag{9.20}$$

where, $\{B_i\}$ is the same as Equation (2.2) and is shown in Figure 2.34, and $\{E_i\}$ includes the last four structuring elements from $\{P_i\}$, that is, the fifth to eighth ones shown in Figure 2.8. In this way, the entire skeletonization process can be controlled automatically; that is, it is repeated until no further changes occur.

To illustrate the effectiveness of this modified algorithm, a simple example is given in Figure 9.18. Figure 9.18a shows the original feature, Figure 9.18b shows the result after applying the skeletionization algorithm without the pruning process, and Figure 9.18c illustrates the result after applying the pruning process.

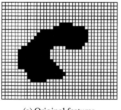 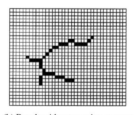 

(a) Original features   (b) Result without pruning process   (c) Result after pruning

FIGURE 9.18 Skeletonization of an irregular area feature produced by the algorithm modified (Reprinted from Su et al., 1998. With permission.).

It should be pointed out here that the algorithms for thin area features are equally applicable to linear features with a certain width, such as rivers and roads represented by double lines. The only different treatment is to close the double lines to form a thin polygon.

9.4.3 PARTIAL COLLAPSE*

In the previous section, a modification was made of a skeletonization algorithm for area-to-line collapse. This algorithm will form the basis for the partial collapse operation discussed in this section.

Partial collapse occurs when a part or some parts of an area feature are long but thin. The main tasks of this operation are (a) to make these long but thin parts a complete collapse and (b) to simplify (the boundary of) the part of the area feature that is not going to be collapsed. Therefore, for the partial collapse operation, the critical issue is how to integrate area-to-line collapse and area simplification operations in a logical manner. Su et al. (1998) described an algorithm (called the Su-Li-Lodwick partial collapse algorithm) based on morphological operators.

The line of thought of the Su-Li-Lodwick partial collapse algorithm is:

1. Areas which are thinner than a certain value (or an objective criterion) are eroded using an erosion operation.
2. Thin parts that are eroded are then replaced by the respective skeletons.
3. The main part of the area that remains after erosion is simplified; that is, its boundary is smoothed.

This basic concept is illustrated in Figure 9.19.

In order to make the procedure described above work, several other processes also need to be involved. Using morphological operators, this process can be implemented by a five-step procedure as follows (Su et al., 1998):

1. Derive the skeleton $Sk(A)$ of area feature A using Equation 9.20.
2. Separate the areal part from the whole feature A using an opening operator with a circular structuring element S_1, the size of which is the width of the linear part in terms of pixel plus 1:

$$C = A \circ S_1 \tag{9.21}$$

3. Simplify the areal feature C with closing and opening operators using Equation 9.22, where S_2 is a critical value to be computed by Equation 9.23 or 9.24 below:

$$Z = (C \bullet S_2) \circ S_2 \tag{9.22}$$

4. Obtain the final result by overlaying $Sk(A)$ and Z.
5. Apply a dilation operator to thicken the skeleton so that it can be represented at a smaller scale.

---

* This section is largely extracted from Su et al., 1998. With permission.

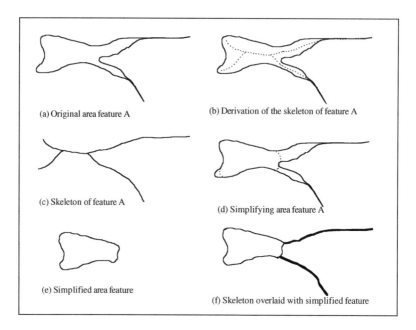

(a) Original area feature A

(b) Derivation of the skeleton of feature A

(c) Skeleton of feature A

(d) Simplifying area feature A

(e) Simplified area feature

(f) Skeleton overlaid with simplified feature

FIGURE 9.19 Process of a partial collapse operation (Reprinted from Su et al., 1998. With permission.).

Figure 9.20 illustrates this partial collapse operation for various scale reduction factors. (In this illustration, pixel size is assumed to 1.5 mm, and D_c is assumed to be 1 pixel size.) The value of S_2, which is a critical value for partial collapse, can be calculated by Equation 9.23:

$$S_2 = \frac{S_{target}}{S_{source}} \times D_c \tag{9.23}$$

where, S_{source} and S_{target} are the scale factors of the source and target data, respectively. D_c is the threshold of the width of an area feature in terms of the number of pixels below which that part of the feature on the target map should be collapsed. Su et al. (1998) call this threshold the *threshold of collapse,* and it is about 0.7 mm, as guided by the natural principle discussed in Chapter 3.

If a symmetric structuring element with the origin at the center is to be used, then the dimension of the structuring element should be an odd number. In this case, Equation 9.23 can be rewritten as follows:

$$S_2 = INT \left(\frac{INT \left(\frac{S_{target}}{S_{source}} \times D_c + 0.5 \right)}{2} \times 2 + 1 \right) \tag{9.24}$$

where *INT* is the integer part of the value. For example, INT(2.8) = 2.

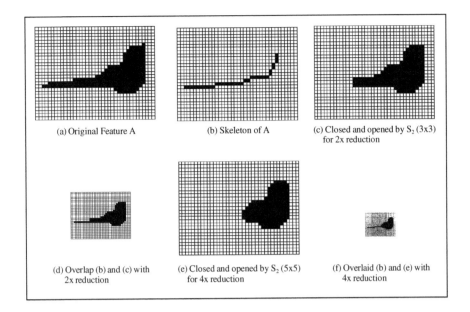

(a) Original Feature A

(b) Skeleton of A

(c) Closed and opened by S_2 (3x3) for 2x reduction

(d) Overlap (b) and (c) with 2x reduction

(e) Closed and opened by S_2 (5x5) for 4x reduction

(f) Overlaid (b) and (e) with 4x reduction

FIGURE 9.20 Morphological transformation for a partial collapse operation by the Su-Li-Lodwick partial collapse algorithm (Reprinted from Su et al., 1998. With permission.).

As illustrated in Figure 3.6 in Chapter 3, the partial collapse can also be automatically produced if the area feature is placed to a set of template. To make the digital recording possible, only those critical cells, near the center of which the feature passes through, should be selected.

9.5 ALGORITHMS FOR AREA ELIMINATION

Elimination of a small area in vector mode is easy, that is, to simply delete the feature from the data set if its size is smaller than a given threshold. However, in raster mode, a rather complicated process is required. In this section some interesting operators will be introduced.

9.5.1 ELIMINATION VIA SEQUENTIAL ERODING USING MONMONIER OPERATORS

In the early 1980s, Monmonier (1983) experimented with digital map generalization in raster mode. He suggested some very interesting ideas for a number of generalization operations such as merging, splitting, elimination, and partitioning of polygons. This section presents his idea for elimination.

In Figure 9.21 a set of five templates is set as the rules, the *Monmonier eroding operators* in this text. In the *eroding operation* the center of a template is moved to

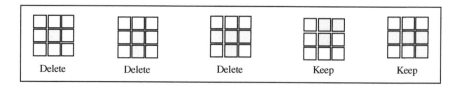

FIGURE 9.21 Monmonier eroding operators.

each pixel in a row and in a column. If the center pixel is white (or zero), then no action is taken. If a pattern similar to the first three templates is found, then the center pixel is deleted. The set of templates will be employed sequentially. The operation continues until no further erosion is possible. Figure 9.22 shows an example of the eroding process (Monmonier, 1983).

9.5.2 ELIMINATION VIA EROSION FOLLOWED BY RESTORATION*

The first line of thought is to erode small features using an erosion operation and then dilate the eroded result. However, the shape of the resultant area features will be oversimplified. To overcome this deficiency, Su et al. (1997a) have developed an algorithm as follows:

1. Apply an erosion operation using a structuring element with an appropriate size.
2. Restore the surviving area features to their original shape and size.
3. Simplify the shape of the restored area features.

Figure 9.23 shows the erosion of a set of area features, with different structuring elements shown in Figure 9.24. It can be seen that fewer and fewer area features can survive from elimination when the size of the structuring element becomes bigger and bigger. This means that the size of the structuring element used in an erosion

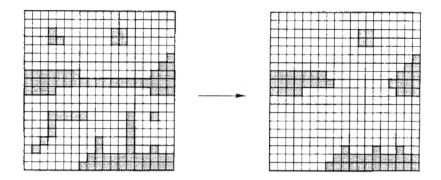

FIGURE 9.22 Elimination of small area features using Monmonier eroding operators (Reprinted from Monmonier, 1993. With permission.).

---

* This section is largely extracted from Su et al., 1997a. With permission.

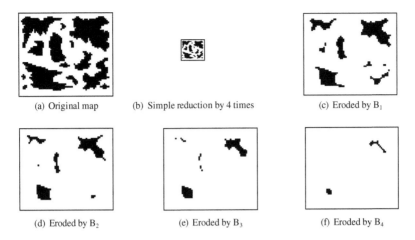

(a) Original map   (b) Simple reduction by 4 times   (c) Eroded by B_1

(d) Eroded by B_2 (e) Eroded by B_3 (f) Eroded by B_4

FIGURE 9.23 Result after applying an erosion process to area features (Reprinted from Su et al., 1997a. With permission.).

operator is a critical element and provides direct control over the output of a single erosion operation. Therefore, it is of crucial importance to determine the appropriate size of a structuring element. To do so, scale is a specific factor to be considered. In other words, the size of the structuring element is dependent on the source scale and the target scale. Its value can be calculated by Equation 9.23. The D_c in this case is the dimension (the distance) at the source scale in terms of the number of pixels below which objects on the source map cannot be further represented. This value is the threshold of representation (equal to or larger than the threshold of perception). For example, a forest area is too small to be represented at a map scale of 1:50 000 if the area is smaller than 100 x 100 m² (Swedish National Land Survey (LMV, 1985)).

The erosion operator can also be repeated several times for a single elimination process. In this case, the number of erosion operations also provides a control over the results of the elimination process. Here, the appropriate size of the structure element to be used is critical.

After application of an erosion operator, some small area features disappear and large features are reduced in size. Those that disappeared are what one wants to eliminate, and those that are reduced in size are what one wants to retain and needs

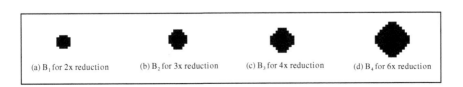

(a) B_1 for 2x reduction   (b) B_2 for 3x reduction   (c) B_3 for 4x reduction   (d) B_4 for 6x reduction

FIGURE 9.24 Structuring elements for various levels of scale reduction.

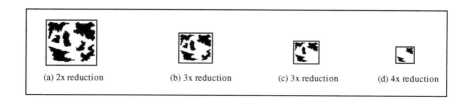

FIGURE 9.25 Area features that survived during erosion process and then were recovered by a restoration process (Reprinted from Su et al., 1997a. With permission.).

to recover. The restoration model developed by Su and Li (1995) (called the Su-Li restoration model here) can be used for such a purpose:

$$R_k = (R_{k-1} \oplus B_{3\times3}) \cap A \qquad (9.25)$$

where A is the original representation (before erosion), k is the round of restoration, R_k is the result obtained after the kth round of restoration, $R_0 = E$ (eroded result), and $B_{3\times3}$ is a squared structuring element with the size 3×3. This is an iteration process that will continue until the following condition holds:

$$R_k = R_{k-1} \qquad (9.26)$$

The results of applying this restoration process to Figure 9.23 are shown in Figure 9.25, with appropriate scale reduction. It can be seen clearly that the area features that are too small to be represented are all eliminated, but those that are large enough to be represented are recovered exactly in terms of both size and shape. However, the shape of these features now appears to be too irregular for representation at a reduced scale. A process needs to be applied to simplify the shape of the area features, and such a process has been discussed in Section 9.3.3.

9.5.3 ELIMINATION BY MODE FILTER

Filters work with a template. In this case the template is similar to the structuring element in mathematical morphology. The size of the template to be employed should be similar to the size of the structuring element discussed in Section 9.5.2. The principle of a *mode filter* is very simple; it makes use of the mode (or dominant value) among many categories as the representative for the central pixel of the template. In data processing using a mode filter, a template (e.g., 3×3) is moved along the image pixels, row by row and column by column. The mode of all the pixels in the template is used as the representative for the position underneath the center of the template. Figure 9.26 shows an example of filtering data with two categories with a 3×3 mode filter.

This is the case with only two categories, that is, either feature or nonfeature. This technique is equally applicable to data with more than two categories. Figure 9.27 shows an example with three categories. Again, a 3×3 filter is employed. In each

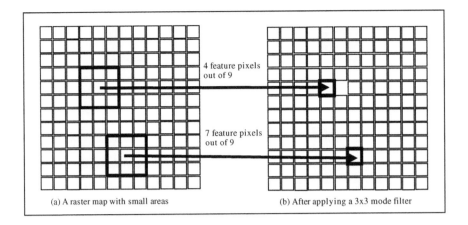

(a) A raster map with small areas   (b) After applying a 3x3 mode filter

FIGURE 9.26 Elimination of small areas by use of a mode filter.

position the mode (or dominant category) with the template is used to represent the central pixel. It is possible to have equal numbers among these categories. If this happens, the neighboring pixels are also considered. Figure 9.27 also shows such a case.

9.5.4 ELIMINATION VIA A CHANGE IN PIXEL SIZE

In the case of the elimination of small area features by a mode filter, the pixel size of the data remains the same. It is also possible to eliminate small area features by a change in pixel size. This process is called resampling.

Figure 9.28 shows two examples, one for a 3×3 template into a new pixel and the other for a 2×2 template into a pixel. In each case the dominant category (i.e., mode) in the template is selected as the representative of this template, and the size of the new pixel is the multiple of the size of old pixels. It is also possible to have

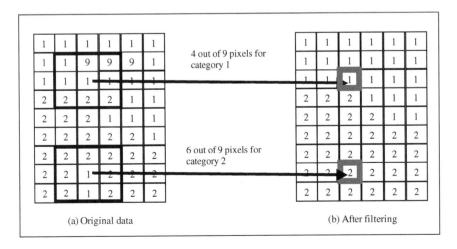

(a) Original data   (b) After filtering

FIGURE 9.27 Elimination of small areas in categorical data by use of a mode filter.

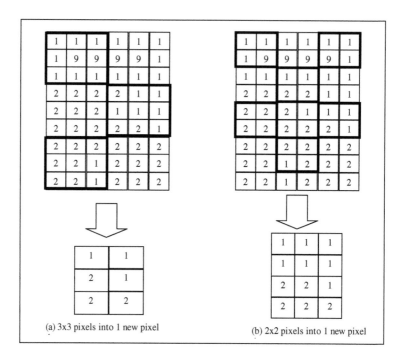

(a) 3x3 pixels into 1 new pixel

(b) 2x2 pixels into 1 new pixel

FIGURE 9.28 Elimination of small area features through a change in pixel size.

the size of the new pixel not exactly equal to a multiple of the size of the old pixels. Figure 9.29 shows that the new pixel size is 1.5 times the old pixel size.

9.5.5 Coarsening as Elimination of an Area Feature

Coarsening is an operation used to eliminate holes or islands in an area feature. Such an operation in vector mode is easy, that is, simply deleting the feature from the data set. However, in raster mode, a rather complicated process is required. Figure 9.30 illustrates one of many possible alternatives. The algorithm can be written as follows:

1. Make the complement of the area feature with islands so as to make the islands and the background as foreground features (i.e., black or 1 pixels).
2. Apply an erosion with a (circular or square) structuring element of an appropriate size or repeatedly apply erosion operations with a small (circular or square) structuring element until the islands are removed.

FIGURE 9.29 Resampling a map of 3 × 3 pixels into 2 × 2 pixels.

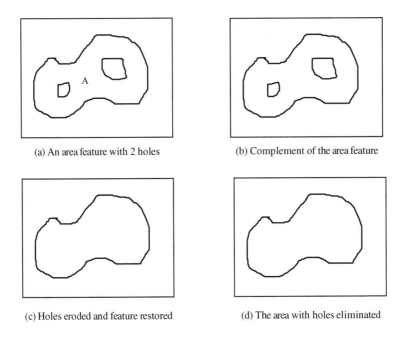

(a) An area feature with 2 holes   (b) Complement of the area feature

(c) Holes eroded and feature restored   (d) The area with holes eliminated

FIGURE 9.30 A procedure for a coarsening operation.

3. Recover the size and shape of the features using the Su–Li restoration model (see Equation 9.25).
4. Make the complement of the restored feature.

Figure 9.30 shows the process. Figure 9.30a shows an area feature A with two holes. Figure 9.30b shows the complement of A, that is, A^C. In this way, A^C becomes foreground features and A becomes background. A^C is then eroded to remove the two holes and then restore the main part that was also eroded at the same time. The restored result is shown in Figure 9.30c. The complement of the restored feature is the final result of coarsening, as shown in Figure 9.30d.

9.6 ALGORITHMS FOR SPLITTING AN AREA FEATURE

Splitting a long and possibly thin area feature is not very convenient in vector mode. However, it is rather easy in raster mode.

9.6.1 SPLITTING VIA SYSTEMATIC ELIMINATION AND ERODING

Monmonier (1983) demonstrated that a long and thin area feature can be split easily and naturally by systematic elimination of rows and columns of the feature image. Figure 9.31 shows an example of this.

It is possible to split such an area feature by Monmonier eroding operators, which has been discussed in Section 9.5.1.

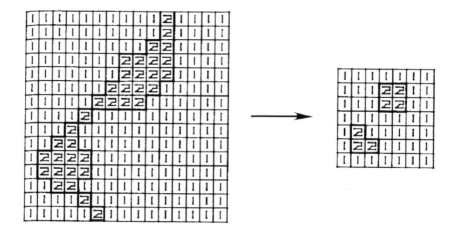

FIGURE 9.31 Splitting through systematic elimination of rows and columns (Reprinted from Monmonier, 1983. With permission.).

9.6.2 SPLITTING VIA MORPHOLOGICAL OPENING

Li (1994) systematically demonstrated the use of morphological opening as an effective solution for splitting an area feature. Such an operation has been defined by Equation (2.14) and clearly demonstrated in Figure 2.18 in Chapter 2, which shows the splitting of a building. Opening is equally applicable to natural area features. Figure 9.32 shows such an example, where "-" refers to pixels eroded after the erosion operation.

9.7 ALGORITHMS FOR EXAGGERATION

The word *exaggeration* has many meanings such as "overstatement" and "amplification." In multi-scale representation, the term *exaggeration* is confusing. Some researchers have used it to refer to the amplification or enlargement of a feature, while others have used it to indicate the enhancement of a characteristic such as the bottleneck of a bay. As shown in Table 1.6 in Chapter 1, three types of exaggeration for area features have been identified: enlargement, widening, and thickening. The widening of the bottleneck of a bay can be achieved by partial modification of the line, which will be discussed in Chapter 11. Only enlargement and thickening will be addressed in this section.

9.7.1 WHOLE EXAGGERATION BY ENLARGEMENT: BUFFERING AND EXPANSION

If the whole area feature is to be exaggerated, terms such as *magnification, enlargement, expansion,* and *amplification* are equally valid.

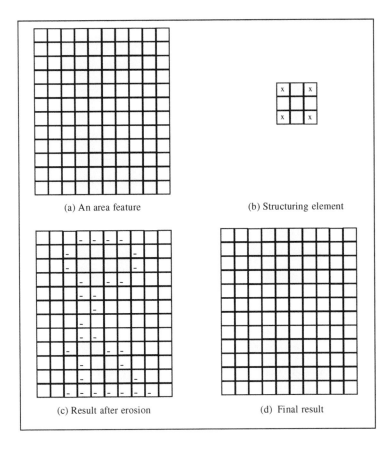

(a) An area feature   (b) Structuring element

(c) Result after erosion   (d) Final result

FIGURE 9.32 Splitting through an opening operation.

In vector mode the buffering operation can be used to expand an area feature in all directions. This is achieved by moving each line segment of the boundary outward in its normal direction. Mathematically, the computation is not very simple.

In raster mode it is very straightforward. The expansion operation of mathematical morphology is the best solution, which is a dilation with a symmetric structuring element such as a circle or a square. Figures 2.15b and 2.15d in Chapter 2 are the result of morphological expansion using a circular or square structuring element.

9.7.2 PARTIAL EXAGGERATION: DIRECTIONAL THICKENING

Partial exaggeration is to expand a feature along a direction. This operation is not too complicated in vector mode. In raster mode a directional thickening could be employed for such a purpose. Figure 9.33 shows a set of eight structuring elements for thickening in eight directions.

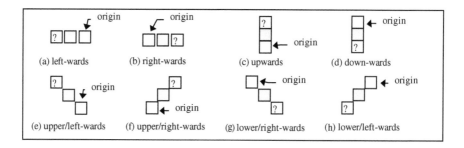

FIGURE 9.33 A set of eight structuring elements for directional thickening (? indicates three options: black, white, or don't care).

The "?" in the structuring element means three options: black pixel (or 1), white pixel (or zero), or "don't care". Figure 9.34 demonstrates the different results obtained with different options for "?" in the structuring element. In this particular example, the structuring element B_1 is an appropriate choice.

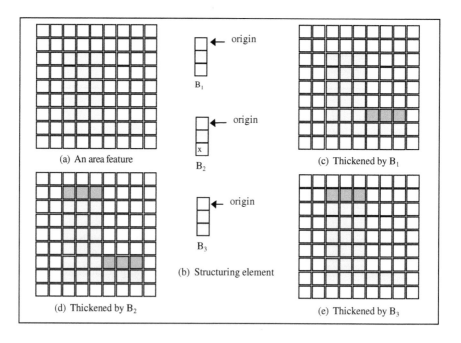

FIGURE 9.34 An example of partial exaggeration by directional thickening.

REFERENCES

Beard, M. K. and Mackaness, W., Generalization operators and supporting structures, *Proceedings Auto-Carto*, held in Baltimore, MD, 1991, pp. 29–45.

Bourke, P., Calculating the Area and Centroid of a Polygon. http://astronomy. swin.edu.au/~pbourke/geometry/polyarea/, 1988.

Chithambaram, R., Beard, K., and Barrera, R., 1991. Skeletonising polygons for map generalisation, *Technical Papers of ACSM-ASPRS Annual Convention, Auto-Carto 10*, Baltimore, MD, March, 6, 44–55, 1991.

de Berg, M., van Kreveld, M., Overmars, M., and Schwarzkopf, O., *Computational Geometry: Algorithms and Applications*, 2nd ed., Springer, Wien, New York, 2000.

DeLucia, A. A., and Black, R., B., A comprehensive approach to automatic feature generalisation, *Proceedings of 13th International Cartographic Conference*, Morelia, Mexico, October, 4, 1987, pp. 169–192.

Freeman, H. and Shapira, R., Determining the minimum-area enclosing rectangle for arbitrary closed curve, *Communication of ACM*, 18, 409–413, 1975.

Graham, R. L., An efficient algorithm for determining the convex hull of a finite planar set, *Information Processing Letter*, 7, 175–180, 1972.

Jones, C. B., Bundy, G. L., and Ware, J. M., Map generalization with a triangulated data structure, *Cartography and Geographic Information Systems*, 22(4), 317–331, 1995.

Li, Z. L., Mathematical morphology in digital generalization of raster map data, *Cartography*, 23(1), 1–10, 1994.

Monmonier, M., Raster-mode area generalisation for land use and land cover maps, *Cartographica*, 20(4), 65–91, 1983.

LMV, *Topographic Map. National Land Survey of Sweden*, Swedish National Land Survey, Gävle, Sweden, 1985. (In Swedish).

Pirzadeh, H., Welcome to Hormoz Pirzadeh's homepage, http://cgm.cs.mcgill.ca/~orm/welcome.html, 1998.

Serra, J., *Image Processing and Mathematical Morphology*. Academic Press, NY, 1982, 610 pp.

Sunday, D., Bounding Containers for Polygons, Polyhedra, and Point Sets (2D & 3D), http://softsurfer.com/Archive/algorithm_0107/algorithm_0107.htm#Minimal%20Rectangle, 2005.

Su, B. and Li, Z. L., An algebraic basis for digital generalisation of area-patches based on morphological techniques, *Cartographic Journal*, 32(2), 148–153, 1995.

Su, B., Li, Z. L., and Lodwick, G., Morphological transformation for the elimination of area features in digital map generalization, *Cartography*, 26(2), 23–30, 1997a.

Su, B., Li, Z. L., and Lodwick, G., Algebraic models for collapse operation in digital map generalization using morphological operators, *Geoinformatica*, 2(4), 359–382, 1998.

Su, B., Li, Z. L., Lodwick, G., and Müller, J. C., Algebraic models for the aggregation of area features based upon morphological operators, *International Journal of Geographical Information Science*, 11(3), 233–246, 1997b.

Toussaint, G. T., Solving geometric problems with the rotating callipers, *Proceedings of IEEE MELECON '83*, Athens, Greece, May 1983, p. A10.02/1-4.

10 Algorithms for Transformations of a Set of Area Features

INTRODUCTION

Chapter 9 presented algorithms for the transformations of individual area features. In this chapter, the focus will shift to the transformation of a set of area features.

10.1 TRANSFORMATION OF A CLASS OF AREA FEATURES: AN OVERVIEW

In this chapter, the first issue to be tackled is the simplification of the shapes of a set of contiguous area features, called a *polygonal network* in this text. This transformation only changes the appearance of individual area features if processed adequately. If it is not processed adequately, the topological relations may be altered as well. This issue will be discussed in Section 10.2.

From a statistical point of view, land use and settlements are two major types of area features. In the transformation of the representation of land use from large scale to small scale, *merging* and *dissolving* are two major operations when the change in scale is not very large. Algorithms for these two operations will be discussed in Section 10.4. However, if the scale change is large, aggregation will become more dominant, for example, areas covered by different types of trees being aggregated into a forest area. Algorithms for aggregation will be covered in Section 10.3.

In the transformation of the representation of settlements from large scale to small scale, many kinds of operations are required such as collapse, aggregation, amalgamation, agglomeration, and typification, depending on the source scale and target scale. For example, when going from 1:10,000 to 1:25,000, the main operation is typification. However, when transforming to 1:50,000, aggregation and amalgamation start to take certain percentage. When further transformed to 1:100,000, aggregation and amalgamation dominate over other operations. If the scale is further reduced to very small scale, a settlement may be collapsed into a point. Algorithms for aggregation and amalgamation will be discussed in Section 10.3, and algorithms for typification will be discussed in Section 10.7.

Algorithms for agglomeration of area features and structural simplification of area patches are discussed in Sections 10.5 and 10.6, respectively.

An important issue related to the transformation of settlements from large scale to small scale is what operation to apply to what context. This lies outside the scope of this text. For better understanding of the contents covered in this chapter, a brief

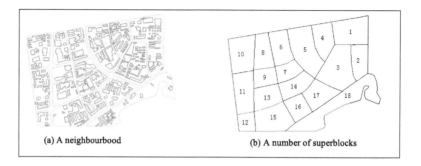

(a) A neighbourbood   (b) A number of superblocks

FIGURE 10.1 Hierarchical structure of settlement in urban morphology (Reprinted from Li et al., 2004).

discussion on this issue will be given in this section. Li et al. (2004) suggest employing the four-level hierarchical structure of a city to deal with the problem, which is developed in urban morphology (see Patricios, 2002), that is, enclave, block, superblock, and neighborhood. That is, a city is first partitioned into many neighborhoods; each neighborbood is then partitioned into many superblocks, each superblock into many blocks, and finally each block into many enclaves. Figure 10.1 shows the partition of a neighborbood into superblocks.

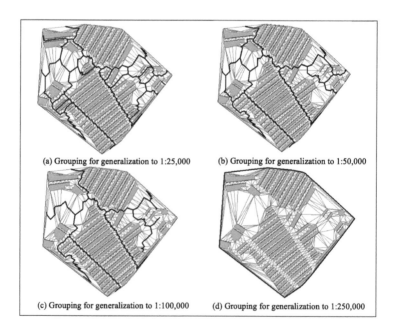

(a) Grouping for generalization to 1:25,000   (b) Grouping for generalization to 1:50,000

(c) Grouping for generalization to 1:100,000   (d) Grouping for generalization to 1:250,000

FIGURE 10.2 Grouping of buildings at 1:10000 scale for generalization to various scales (Reprinted from Li et al., 2004).

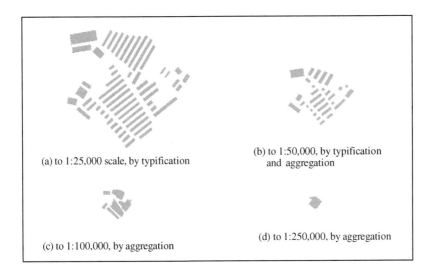

(a) to 1:25,000 scale, by typification

(b) to 1:50,000, by typification and aggregation

(c) to 1:100,000, by aggregation

(d) to 1:250,000, by aggregation

FIGURE 10.3 Transformation of grouped buildings to various scales (Reprinted from Li et al., 2004).

Within each superblock, buildings are put together into groups according to their topological adjacency, separation, and alignment. Two adjacent groups may be merged if they have the same type of alignment. Figure 10.2 shows the grouping of buildings into enclaves and blocks for further processing, and Figure 10.3 shows the transformation of the grouped buildings to a different scale by different operations.

10.2 ALGORITHMS FOR SIMPLIFICATION OF THE SHAPE OF A POLYGONAL NETWORK

Polygonal network here refers to the tessellation of two-dimensional (2-D) space by contiguous but nonoverlapping polygons of possibly irregular shape. This term is an analogy of *triangular network,* which is popularly used in spatial data processing and was discussed in Chapter 2.

The simplification of the shape of a polygonal network can be carried out through two different approaches: decomposition-based or whole-based.

10.2.1 DECOMPOSITION-BASED SIMPLIFICATION OF A POLYGONAL NETWORK

The decomposition-based approach consists of the following three steps:

1. Decompose each polygon into a set of polylines (arcs).
2. Simplify each polyline separately.
3. Assemble the separately simplified polylines back into a polygonal network.

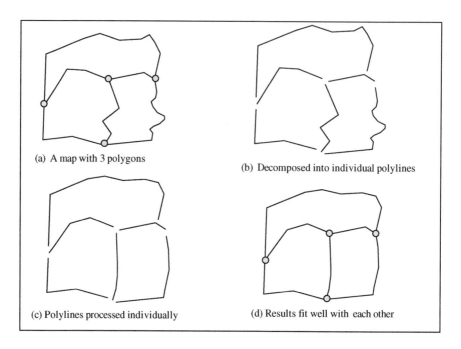

(a) A map with 3 polygons

(b) Decomposed into individual polylines

(c) Polylines processed individually

(d) Results fit well with each other

FIGURE 10.4 Decomposition-based approach for simplifying the shape of a polygonal network.

The topological relations between different polygons are established using the conventional topology, which is encoded into three tables expressing the relationships between vertices, polylines (arcs), and polygons. Figure 10.4 shows such a process. In this approach the algorithms for individual lines described in Chapters 5 to 7 may be used, so no further discussion is required here. Figure 10.5 is an example of a decomposition-based simplification of the polygonal network of China with the Li–Openshaw algorithm.

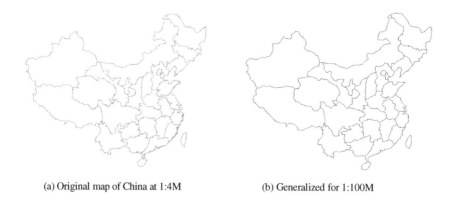

(a) Original map of China at 1:4M

(b) Generalized for 1:100M

FIGURE 10.5 Decomposition-based simplification of polygonal network with the Li–Openshaw algorithm.

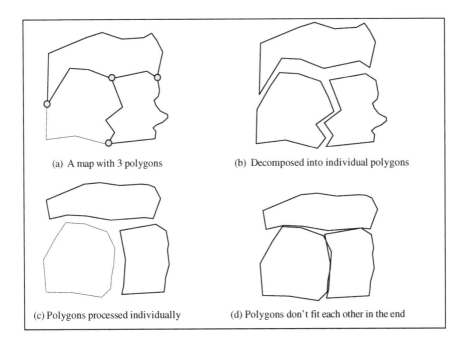

(a) A map with 3 polygons   (b) Decomposed into individual polygons

(c) Polygons processed individually   (d) Polygons don't fit each other in the end

FIGURE 10.6 Whole-based approach for simplifying the shape of a polygonal network.

10.2.2 WHOLE-BASED SIMPLIFICATION OF A POLYGONAL NETWORK

It has been recognized that the work involved in the building and maintenance of the topology is huge. As a result, in spatial information systems the trend is to move away from traditional encoded topology to on-the-fly topology. In such a system, an individual polygon in a whole is recorded and maintained as an entity, and the topological relations between polygons (see Chapter 1 for more details) are established whenever there is a need. Such a system calls for simplifying individual polygons separately, leading to the whole-based approach.

The whole-based approach consists of the following two steps:

1. Simplify each polygon separately.
2. Form a polygonal network from the separately simplified polygons.

With this approach, the polygons after simplification may not fit each other if special care is not taken. Figure 10.6 shows that the intersections may be lost.

Therefore, we suggest using the Li–Openshaw algorithm in raster mode to simplify individual polygons in a polygonal network because the intersections of polygonal boundaries can be preserved automatically by this algorithm. As shown in Figure 10.7 an intersection is recorded as a cell (pixel), no matter how many

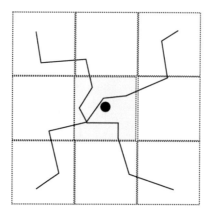

FIGURE 10.7 Intersections preserved automatically by raster mode Li–Openshaw algorithm.

vertices (points) are located and wherever they are located within this cell. It is also possible to use the Li–Openshaw algorithm as suggested in Section 7.4.5 for non-natural area futures. Figure 10.8 shows an example of whole-based simplification of the polygonal network of China with the Li–Openshaw algorithm in raster mode.

10.3 ALGORITHMS FOR COMBINING AREA FEATURES: AGGREGATION AND AMALGAMATION

From a geometric point of view, the grouping of area features in aggregation and amalgamation is the same if only a single layer is considered. Therefore, a more neutral term, *combination*, is used for the grouping of area features to form a new area feature. After combination, a *shape refinement* process should be applied to

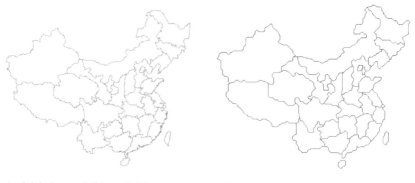

(a) Original map of China at 1:4M   (b) Generalized for 1:100M

FIGURE 10.8 Whole-based simplification of a polygon network by the Li–Openshaw algorithm.

simplify the outline of the combined area feature. As the shape refinement was discussed in Chapter 9, this section only addresses the combination process.

Similar to shape refinement, area combination can also be carried using boundary-based algorithms or region-based algorithms.

10.3.1 BOUNDARY-BASED COMBINATION VIA EQUAL-SPANNING POLYGONS

In some systems dealing with spatial data, the *walking algorithm* based on a distance criterion has been implemented for the combination of area features. It can be written as follows:

1. Select a convenient point as starting point, for example, the point with minimum Y coordinates.
2. Draw a circle with the given distance as its radius from the starting point.
3. Select the intersection (among many possibilities) that makes the largest right-hand turn from the starting point as a boundary point for the combined area.
4. Make the selected point the new starting point.
5. Repeat steps 2 to 4 until the last circle contains the initial starting point.

The result is an irregular polygon with equal-spanning, called an *equal-spanning polygon* here. Figure 10.9 shows the principle of such an algorithm. Figure 10.9a shows the combination of irregular polygons, and Figure 10.9b shows the combination of regular polygons. The results are not very sensible.

10.3.2 BOUNDARY-BASED COMBINATION VIA CONVEX HULLS

An alternative is the use of a *convex hull*, which defines the area of interest that is the smallest convex polygon containing all areas. A number of algorithms are available for constructing the convex hull of a set of points on a 2D plane such as

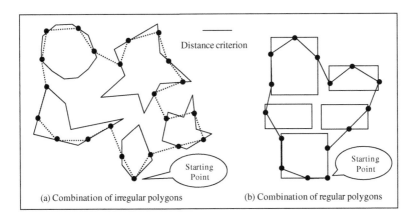

(a) Combination of irregular polygons   (b) Combination of regular polygons

FIGURE 10.9 Combination of area features by the walking algorithm.

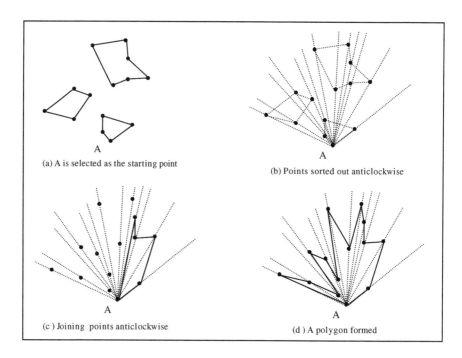

(a) A is selected as the starting point

(b) Points sorted out anticlockwise

(c) Joining points anticlockwise

(d) A polygon formed

FIGURE 10.10 Polygon formation from a set of points by the Graham scan algorithm.

Graham's scan, Jarvis' march (gift-wrapping), and quick hull (Gosper, 1998; O'Rourke, 1993). These algorithms can also be used to construct a convex hull for area features by considering all the boundary points of all polygons as random points.

Some ideas of the *Graham scan algorithm* were partially presented in Section 9.2, where the formation of a convex hull from an irregular polygon was discussed using the so-called three-coins algorithm. The only part left to be discussed here is the formation of a polygon from a set of randomly distributed points. In the Graham scan algorithm a radial sweep method is employed. It first selects a starting point, which must be on the boundary of the convex hull to be formed. Therefore, one of the extreme points, for example, Y_{min}, can be used. Then, each point bearing of the line joining the starting point and the point is computed, and all the points are sorted out counterclockwise according to the bearing. A polygon can then be formed by connecting these points in sequence. Figure 10.10 shows the process. The three-coins method discussed in Section 9.2 is then applied to form a convex hull from the polygon.

The *quick hull algorithm* works in a way similar to the point-reduction algorithms based on perpendicular distance. Figure 10.11 shows the formation of a convex hull by this algorithm. The algorithm can be written as follows:

1. Find two extreme points, for example, one with X_{min} and the other with X_{max}, to form an initial baseline.
2. Find the point with the maximum perpendicular distance to the baseline from one side.

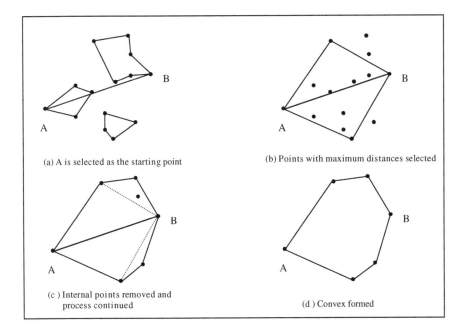

(a) A is selected as the starting point

(b) Points with maximum distances selected

(c) Internal points removed and process continued

(d) Convex formed

FIGURE 10.11 Convex hull formation by the quick hull algorithm.

3. Form a triangle with the selected point and the baseline.
4. Delete all points within this triangle.
5. Use each of the two sides of the triangle as a new baseline.
6. Repeat steps 2 to 5 until all the points on one side have been used or removed.
7. Perform the same process for the other side of the initial baseline.

The *gift-wrapping algorithm* is a very simple and popular algorithm. It is illustrated in Figure 10.12 and can be written as follows:

1. Find the point with the minimum Y coordinate to use as a starting point.
2. Find the point to form a baseline joining this point and the starting point by scanning through all the points to make all other points lie to the left of this base line. That is, this current point under consideration makes the largest right-hand turn from the starting point.
3. Let the current point become the new starting point and repeat step 2 until the convex polygon is closed.

An alternative is to form a convex hull for each polygon and then to combine convex hulls. The Jarvis gift-wrapping algorithm can still be applied for such a purpose. Perhaps the more efficient way is to find the two bridges between each pair

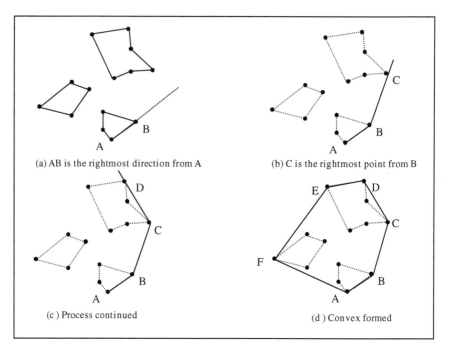

(a) AB is the rightmost direction from A

(b) C is the rightmost point from B

(c) Process continued

(d) Convex formed

FIGURE 10.12 Jarvis's march (gift-wrapping) algorithm for the formation of a convex hull.

of convex hulls (see Figure 10.13). Suppose convex hull P consists of m points as $\{P_1, P_2, ..., P_m\}$ and convex hull Q consists of n points as $\{P_1, P_2, ..., P_m\}$. A pair of points (P_i, Q_j) forms a bridge between two convex hulls if P_{i-1}, P_{i+1}, Q_{j-1}, and Q_{j+1} all lie to the same side of the line joining (P_i, Q_j). Pirzadeh (1998) found that the Toussaint rotating callipers algorithm has two advantages over others: (a) no backtracking is needed and (b) the polygons can intersect. (Other algorithms require the polygons to be disjoint). Further discussion of this lies outside the scope of this text. For those who are interested in this topic, the information provided on the Web page by Pirzadeh (1998) provides a good introduction.

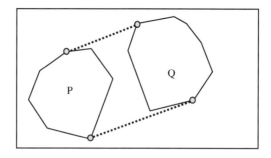

FIGURE 10.13 Two bridges required to form a new convex hull from a pair of convex hulls.

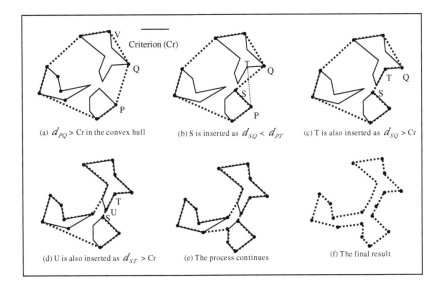

(a) $d_{PQ} >$ Cr in the convex hull

(b) S is inserted as $d_{SQ} < d_{PT}$

(c) T is also inserted as $d_{SQ} >$ Cr

(d) U is also inserted as $d_{ST} >$ Cr

(e) The process continues

(f) The final result

FIGURE 10.14 Constrained hull for combining area features.

10.3.3 BOUNDARY-BASED COMBINATION VIA CONSTRAINED HULLS

The use of convex hulls for combining area features is oversimplified in some cases. Figure 10.14a is an example in which the proportion of open space in the convex hull is much larger than that of the features. Therefore, the hull may be modified, leading to a constrained hull.

One simple line of thought is to set a distance criterion for the sides of the hull. Such a distance can be regarded as the Cantor's ε-chain in computational topology or the Euclidean minimal spanning in Euclidean or the $1/\alpha$ in alpha-shape (α-shape) (Edelsbrunner et al., 1983). Figure 10.14 shows the "stripping" process. Figure 10.14b shows that point S is added because the distance between P and Q, d_{PQ}, is larger than the criterion and because $d_{SQ} < d_{PT}$. This is not enough, as d_{SQ} is still larger than the criterion, so point T is also added. After another check, it is found that d_{ST} is still larger than the criterion, so point U is also inserted. The process continues to check each side of the hull. If the side of the hull has a length larger than the criterion formed by two points (e.g., Q and V) from the same polygon, then the point with numbering in the middle of these points is inserted.

In the stripping process for the constrained hull, the selection of a distance criterion is critical. If the distance is too short, one may fail to complete the process. Therefore, an automated process is very desirable. To automate such a process, two points with the shortest distance between two polygons may be used as connecting points. However, this criterion is not without problems. It can also be found that points U, S, and W in Figure 10.15b show the three shortest distances among the three polygons. The result of using the shortest distance as a criterion is not acceptable because the same point has been used to connect with more than one polygon. Therefore, another

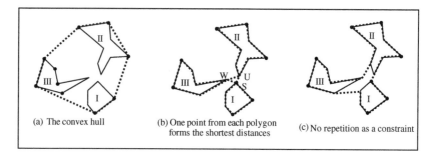

(a) The convex hull

(b) One point from each polygon forms the shortest distances

(c) No repetition as a constraint

FIGURE 10.15 Using the shortest distance as a connection criterion for automating the formation of a constrained hull.

constraint should be set to forbid the repeated use of the same point as a connector to other polygons. Such a constraint improves the result, as shown in Figure 10.15c. Figure 10.16 is an example showing the three types of hulls for combining buildings.

10.3.4 REGION-BASED COMBINATION VIA GAP BRIDGING

Gap bridging is a technique proposed by Monmonier (1983) for the generalization of raster maps. This method "requires that the computer scan along the rows, columns and principal diagonal directions of the raster grid." In the case of areas with a single type of attribute, the procedures can be written as follows:

1. Define the size of the gap (in terms of number of pixels) to be filled up.
2. Search along rows, columns, and diagonal directions for feature (black or 1) pixels. In each search, when a nonfeature pixel is encountered, extend the search outward for a limited distance, that is, the specified gap size.
3. Change the nonfeature pixels into feature pixels if a feature pixel is found after the extended search.
4. Repeat steps 2 and 3 until the result is stable.

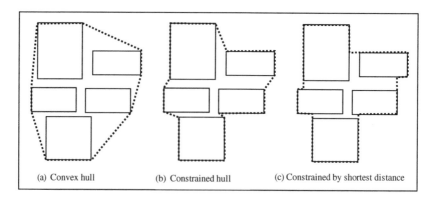

(a) Convex hull

(b) Constrained hull

(c) Constrained by shortest distance

FIGURE 10.16 Using the shortest distance as a connection criterion for the formation of a constrained hull from buildings.

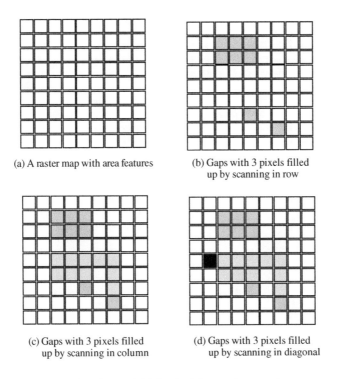

(a) A raster map with area features

(b) Gaps with 3 pixels filled
up by scanning in row

(c) Gaps with 3 pixels filled
up by scanning in column

(d) Gaps with 3 pixels filled
up by scanning in diagonal

FIGURE 10.17 Combination via gap bridging by Monmonier scanning.

Figure 10.17 is an example of a single type of feature only. If there is more than one type of attribute, the procedure should be changed. In this case the term *non-feature* should be changed to *a feature with lower ranking*. Accordingly, *feature* should be changed to *high-ranking feature*.

10.3.5 REGION-BASED MORPHOLOGICAL COMBINATION

Through an analysis of the characteristics of area combination, Su et al. (1997) suggested a simple algebraic model (called the *SLLM combination model* here) as follows:

$$C = (A \oplus B_1) \ominus B_2 \tag{10.1}$$

where A is the representation (image) showing the original features and B_1 and B_2 are the two structuring elements. When $B_1 = B_2$, Equation 10.1 becomes the closing operator. The success of applying this model to area combination depends on the proper size and shape of structuring elements B_1 and B_2. Equations 9.23 and 9.24 in Chapter 9 can be used as a guideline. However, the D_c in Equation 9.23 should be replaced by the *threshold of separation*, which is approximately 0.2 mm in terms of map distance at the target map scale.

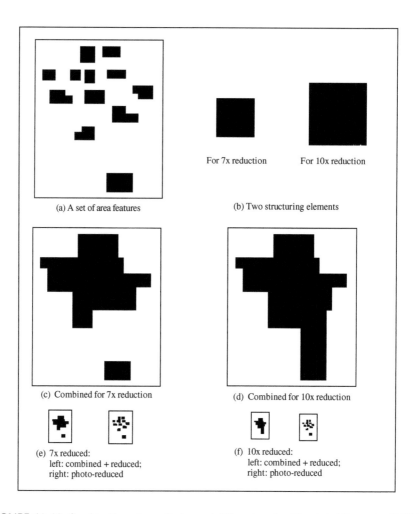

(a) A set of area features

(b) Two structuring elements

For 7x reduction   For 10x reduction

(c) Combined for 7x reduction

(d) Combined for 10x reduction

(e) 7x reduced:
   left: combined + reduced;
   right: photo-reduced

(f) 10x reduced:
   left: combined + reduced;
   right: photo-reduced

FIGURE 10.18 Combination of area features at different scales (Reprinted Su et al., 1997b).

Figure 10.18 shows the combination of buildings using this model for two different scales, one for scale reduction by 7 times and the other by 10 times. The results are also compared with combination by simple photo-reduction. It is clear that the combined results are very reasonable.

The global and general shape of area features need to be kept after the combination operation. To do so, the shape of the structuring element should be kept in accordance with the original shape of the area features. In general, it is suggested that rectangular structuring elements be used for rectangular features and circular structuring elements be used for curved features (i.e., those with natural shapes) (Su et al., 1997). The examples in Figures 10.19 and 10.20 show the combination of various area features with different types of structuring elements.

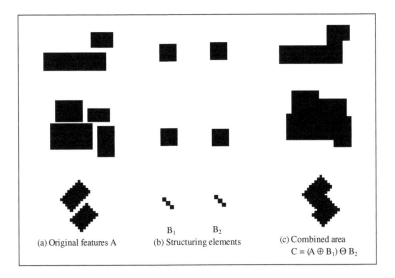

FIGURE 10.19 Combination of rectangular features (Reprinted from Su et al., 1997b).

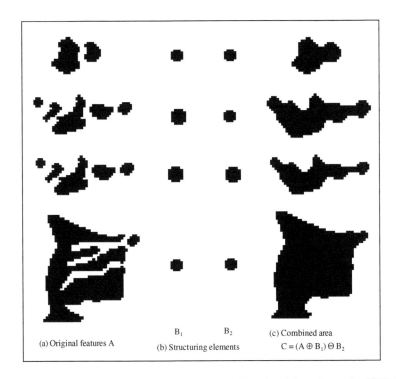

FIGURE 10.20 Combination of curved area features (Reprinted from Su et al., 1997b).

10.4 ALGORITHMS FOR MERGING AND DISSOLVING AREA FEATURES

Merging and dissolving are similar and thus are discussed together in this section.

10.4.1 MERGE VIA A UNION OPERATION

Merge refers to different operations. Sometimes *merge* is used as a synonym of *aggregation.* In this book *merge* refers to the combination of two immediate neighboring area features, but the one which is merged into another loses its identity. In some of the literature, *merge* is also termed *fusion.*

Merging is an easy operation. The decision of which larger area a small area will be merged into can be made based on one of the following three criteria:

1. The area with largest size (Figure 10.21b)
2. The area with longest common boundary (Figure 10.21c).
3. The area with highest importance (Figure 10.21d)

In the case of Figure 10.21b, the merge operation can be expressed as follows:

$$A_{New} = A \cup C$$
$$A \Leftarrow A_{New}$$
(10.2)

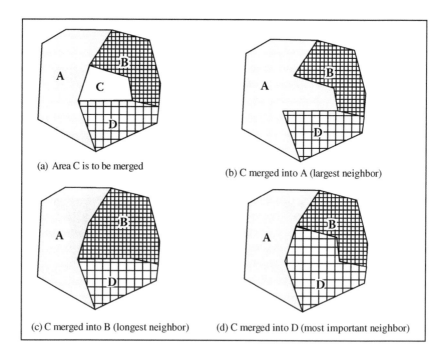

(a) Area C is to be merged

(b) C merged into A (largest neighbor)

(c) C merged into B (longest neighbor)

(d) C merged into D (most important neighbor)

FIGURE 10.21 Merging an area feature into another (one of its neighbors).

where C is an area to merged, A is an area into which C is merged, and A_{New} is the union of A and C. In a merge operation, no new feature is created, so A_{New} is assigned to ($\Leftarrow$) A in the end.

10.4.2 DISSOLVE VIA SPLIT AND MERGE

One simple solution for the splitting process is to make use of the centroid of the small area to be dissolved. With this centroid, the small area is divided into small pieces, and each of its neighboring area features will get a piece. The *dissolve* process via *split and merge* is illustrated in Figure 10.22, and the algorithm could be written as follows:

1. Find the intersections (e.g., T_1, T_2, and T_3 in Figure 10.22b) among the area features, (e.g., A, B, and C in Figure 10.22a).
2. Compute the centroid of the small area to be dissolved (e.g., area feature C in Figure 10.22a).
3. Delete this small area feature to be dissolved.
4. Connect the intersections to the centroid so that each of the neighboring area features takes a part from the dissolved area (Figure 10.22c).

Monmonier (1983) noticed that merge and dissolve are two operations that can be easily implemented in raster. The process is illustrated in Figure 10.23, and the algorithm can be written as follows:

1. Expand the area features to be dissolved by one pixel at each side, for example, C in Figure 10.23a.
2. Intersect the expanded area feature, C, with the original map (Figure 10.23b).
3. Subtract the small area feature to be dissolved from the expanded area to obtain a string of pixels of the neighboring area features (Figure 10.23c).
4. Contract the small area to be dissolved from all directions, pixel by pixel (Figure 10.23d).

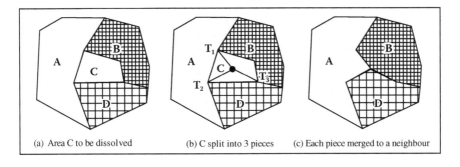

(a) Area C to be dissolved   (b) C split into 3 pieces   (c) Each piece merged to a neighbour

FIGURE 10.22 Dissolving a small area in vector mode.

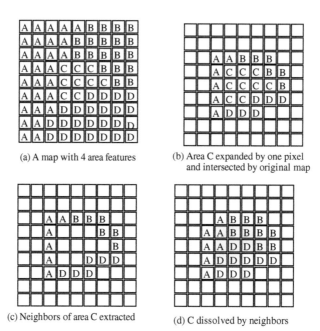

(a) A map with 4 area features

(b) Area C expanded by one pixel and intersected by original map

(c) Neighbors of area C extracted

(d) C dissolved by neighbors

FIGURE 10.23 Dissolving a small area feature in raster mode.

It should be noted here that the method used to extract the string of pixels of the neighboring area features was developed by Schylberg (1993).

10.5 ALGORITHMS FOR AGGLOMERATION OF AREA FEATURES

In some cases the area features are separated by a barrier with a certain width or strips of open space (Figure 10.24a). If the scale is reduced, then the barriers (or strips of open space) become too narrow to be represented (Figure 10.24b), but the nature of the spatial distribution needs to be maintained. In this case it is not appropriate to apply the aggregation operation to combine all of these area features together. A compromise is to represent the barriers (or strips of open space) by single lines (Figure 10.24f).

This is a case of collapse. That is, the barriers are to be collapsed into lines. Therefore, the algorithms discussed in Section 9.5.2 in Chapter 9 can all be used for such a purpose. However, if the barriers are strips of open space, then an explicit representation can be obtained by making the complement of the area features (enclosed by the minimum bounding rectangle) as shown in Figure 10.14c.

A procedure for agglomeration can be written as follows:

1. Produce the complement of the representation so as to make the barriers into features.
2. Derive the skeleton for each barrier feature.
3. Use the skeletons to form the new boundaries of area features.

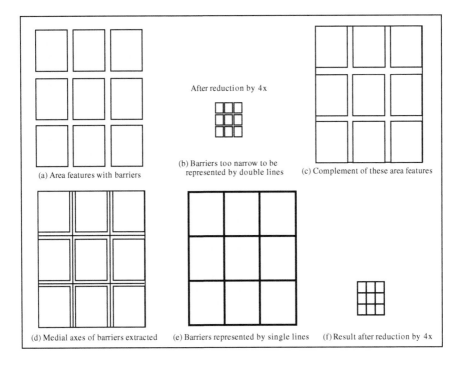

FIGURE 10.24 The process of agglomeration of area features.

10.6 ALGORITHMS FOR STRUCTURAL SIMPLIFICATION OF AREA PATCHES*

In some situations small area features with the same thematic meaning are spread over a certain area. Figure 10.25a is an example (Muller and Wang, 1992) of this, which is sometimes called *area patches*. These small area features could be forest coverage, land use, islands, water bodies, and so on. When map scale is reduced, some area features will be too small to be shown, some will touch, and other may coalesce. Therefore, the structure of the area-patches needs to be simplified.

10.6.1 VECTOR-BASED STRUCTURAL SIMPLIFICATION

Muller and Wang (1992) have considered these problems and produced an algorithm for the structural simplification of area patches. They recommend three general requirements:

1. Only a subset of original patches is preserved, and those selected will be exaggerated in size.
2. Elements of the counter of the convex hull delimiting the original distribution of the patches must be recognizable after the transformation.
3. The structure of the spatial distribution of the patches must be preserved.

---

* This section is largely extracted from Su and Li, 1995. With permission.

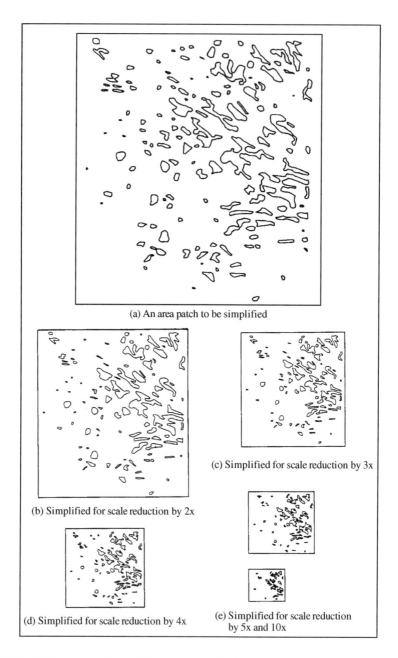

(a) An area patch to be simplified

(b) Simplified for scale reduction by 2x

(c) Simplified for scale reduction by 3x

(d) Simplified for scale reduction by 4x

(e) Simplified for scale reduction by 5x and 10x

FIGURE 10.25 Area-patch simplification done with the Muller–Wang procedure (Reprinted from Muller and Wang, 1992. With permission.).

After considering these requirements, Muller and Wang (1992) then developed a procedure called the *Muller–Wang procedure,* as follows:

1. Rank all patches by size and calculate the cumulative area.
2. Set a threshold for determining whether a patch should be expanded or contracted.
3. Perform expansion for areas whose sizes are larger than the threshold and contraction for those whose sizes are smaller than the threshold (unless the area is far away from or very close to another).
4. Perform elimination of area features that are too small after contraction.
5. Reselect a few eliminated area features in regions where area features are sparsely distributed.
6. Expand all areas features that are retained and/or reselected.
7. Merge overlapping or touching area features.
8. Displace coalesced features and smooth the contours of the patches.
9. Verify topological integrity.
10. Smooth the contours of the patches.

10.6.2 RASTER-BASED STRUCTURAL SIMPLIFICATION

Su and Li (1995) have proposed an algebraic model, the *Su–Li procedure*, to mimic the procedure developed by Muller and Wang (1992). The *Su–Li procedure* empolys the operators developed in mathematical morphology. The procedure is as follows:

1. Determine the size of the structuring elements, which is identical to that used for the elimination of area features discussed in Section 9.5.2.
2. Apply erosion to all area features to eliminate area features that are too small to be represented.
3. Restore the surviving area features (see the restoration algorithm in Section 9.5.2).
4. Apply dilation to all the area features after restoration.
5. Apply erosion to the dilated data (image), if desirable.
6. Apply a postprocessing procedure, if desirable.

Su and Li (1995) have suggested that a smaller structuring element could be used in step 5. In this way, the need to displace coalesced features, as in step 8 in the Muller–Wang procedure, can be avoided. As a consequence, there will be no need to verify the topological integrity.

The postprocessing procedure is composed of the following steps:

1. Reselect some eliminated area patches, called "lucky losers," to give a balanced spatial distribution.
2. Exaggerate the size of those lucky losers so that they are large enough to be represented at a smaller scale.
3. Exaggerate the size of the remaining area patches if desirable.

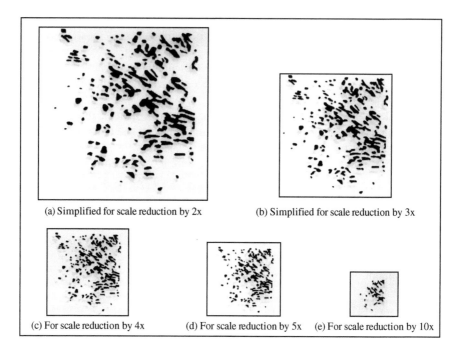

(a) Simplified for scale reduction by 2x   (b) Simplified for scale reduction by 3x

(c) For scale reduction by 4x   (d) For scale reduction by 5x   (e) For scale reduction by 10x

FIGURE 10.26 Structural simplification of area patches with the Su–Li procedure (Reprinted from Su and Li, 1995. With permission.).

Figure 10.26 shows the structural simplification of the set of area patches, by Muller and Wang (1992), now using the Su–Li procedure. The results are quite similar.

10.7 ALGORITHMS FOR TYPIFICATION OF AREA FEATURES

Typification means the reduction of the number of features while preserving the appropriate characteristics of the pattern. In a sense, the structural simplification of area patches is sometimes a kind of typification. However, in most of the literature, typification is applied to buildings.

Buildings may be distributed regularly and irregularly. It would be easier if they were all distributed exactly along rows and columns, but this is not always the case and thus some special treatment must be made during the typification process.

10.7.1 TYPIFICATION OF ALIGNED AREA FEATURES

Regnauld (2001) has published a lengthy paper describing the typification of buildings. The process is illustrated in Figure 10.27a, and the algorithm, called the *Regnauld typification algorithm* here, can be written as follows:

1. Analysis and partitioning: To partition buildings into groups. This process uses the minimum spanning tree to build a proximity graph (Figure 10.27a). This graph is then segmented into building groups by breaking the link between two vertices if some distance-related criteria are met (Figure 10.27b);

2. Harmonization of building sizes: To ensure that the buildings are large enough for scale reduction. The increase in the minimum size should be mirrored by changes in size for other buildings.

3. First positioning: To place some buildings to ensure separation and the preservation of the relative positions between groups. All pairs of buildings at the end of an eliminated edge are placed at their original position (Figure 10.27c).

4. Second positioning (local grouping): To typify each group. During this process, important buildings are positioned and then the gaps are then filled in. Buildings are displaced along an imaginary line, that is, the line connecting the centroids of buildings linked by the initial minimum spanning tree (Figure 10.27d).

Li et al. (2004) have also discussed the typification of buildings after the grouping process, called the LYAC typification algorithm. They made use of the attribute information of the group, such as the sum of the building's area, the mean separation, and the standard deviation of the separation of buildings, to decide when typification should be applied. After a decision is made, the typification process is as follows:

1. Calculate the sum length of the linearly scattered alignment.
2. Let the number of the resultant buildings in the group be equal to M.
3. Calculate the separation between and the length and width of the buildings.

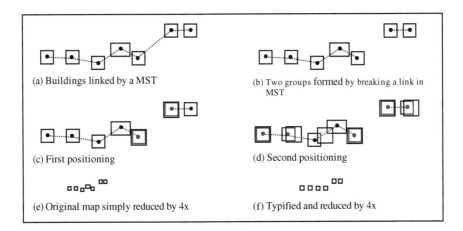

(a) Buildings linked by a MST

(b) Two groups formed by breaking a link in MST

(c) First positioning

(d) Second positioning

(e) Original map simply reduced by 4x

(f) Typified and reduced by 4x

FIGURE 10.27 Building typification by the Regnauld algorithm.

4. Let $N = M$ and go to step 6 if the separation is greater than or equal to the separation threshold (i.e., the minimum distance between two buildings that is required for clarity, generally 0.5 mm) and the orientation of major axes are consistent with their original ones.
5. Make the number (M) of resultant buildings in the group one less than the previous value and go to step 3.
6. Arrange the first and last buildings at their original positions to ensure the preservation of the group structure (also see Figure 10.27d).
7. Fill in the other ($N - 2$) buildings according to the computed separation.

10.7.2 TYPIFICATION OF IRREGULARLY DISTRIBUTED AREA FEATURES

Basaraner and Selcuk (2004) have described the typification of irregularly distributed building features using a Laser-Scan (2001) system. The Basaraner–Selcuk typification algorithm is as follows:

1. Collapse building clusters into point clusters.
2. Group building points into clusters using hierarchical clustering by dendrogram (which is built by repeatedly finding the two closest buildings [unless the distance is larger than a given threshold], adding them to the tree, and creating a new node to represent this new cluster. This new node is added back to the list for further clustering).
3. Use a new building to replace the two clustered old buildings, with the center, orientation, length, and width all being the average of the old two.

Figure 10.28 shows an example of such a typification. Li et al. (2005) proposed a similar algorithm for web mapping as follows.

1. Collapse building clusters into point clusters.
2. Group building points into clusters.
3. Determine the number of buildings to be retained in each cluster according to the Radical Law.

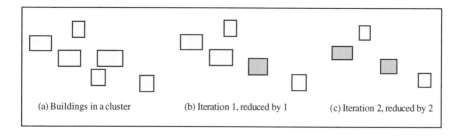

(a) Buildings in a cluster   (b) Iteration 1, reduced by 1   (c) Iteration 2, reduced by 2

FIGURE 10.28 Typification of irregularly distributed buildings by iterative elimination.

4. Determine the representation of new buildings using an iterative procedure. Each time a new building is created, replace the two closest old buildings, with the center, orientation, length, and width all being the average of the old two. Only if the size difference is very large, should the bigger area be retained and the small area eliminated. The number of buildings in each cluster is then reduced by one. If the number of buildings in the cluster is still more than the desired number, repeat step 4.

5. Harmonize the size of buildings to ensure that the minimal size is met and the size differences between buildings are relatively preserved (see step 2 of the Regnauld algorithm described in Section 10.7.1).

REFERENCES

Basaraner, M. and Selcuk, M., An attempt to automate generalization of buildings and settlement areas in topographic maps, in *International Archives of Photogrammetry and Remote Sensing*, (Proceedings of XXth International Congress for Photogrammetry and Remote Sensing, Istanbul), *XXXV(Part B4), 2004. (CD-Rom)*.

Edelsbrunner, H. and Mücke, E. P., Three-dimensional alpha shapes, *ACM Transaction in Graphics*, 13(1), 43–72, 1994.

Edelsbrunner, H., Kirkpatrick, D. G., and Seidel, R., On the shape of a set of points in the plane, *IEEE Transaction in Information Theory*, IT-29, 551–559, 1983.

Gosper, J..J., 2D Convex Hulls, http://www.brunel.ac.uk/~castjjg/java/mscthesis/convexhull/, 1998.

Laser-Scan, *The Gothic Module Reference Manual*, Laser-Scan Ltd. Cambridge, UK, 2001.

Li, H. S., Guo, Q. S., and Liu, J. P., Rapid algorithm of building typification in web mapping, *International Archives of Photogrammetry and Remote Sensing*, XXXVI(/W), 145–150, 2005.

Li, Z. L. and Openshaw, S., Algorithms for automated line generalisation based on a natural principle of objective generalisation, *International Journal of Geographical Information Systems*, 6(5), 373–389, 1992.

Li, Z. L., Yan, H., Ai, T., and Chen, J., Automated building generalization based on urban morphology and gestalt theory, *Journal of Geographical Information Science*, 18(5), 513–534, 2004.

Mandal, D. P. and Murthy, C. A., Selection of alpha for alpha-hull in R^2, *Pattern Recognition*, 30(10), 1759–1767, 1997.

Monmonier, M., Raster-mode area generalisation for land use and land cover maps, *Cartographica*, 20(4), 65–91, 1983.

Muller, J.-C. and Wang, Z., Area-patch generalization: a competitive approach, *Cartographic Journal*, 29(2), 137–144, 1992.

O'Rourke, J., *Computational Geometry in C*, Cambridge Press, Cambridge, UK, 1993.

Patricios, N. N., Urban design principles of the original neighbourhood concepts, *Urban Morphology*, 6, 21–32, 2002.

Pirzadeh, H., Welcome to Merging Convex Hulls, http://cgm.cs.mcgill.ca/~orm/mergech.html, 1998.

Regnauld, N., Context building typification in automated map generalization, *Algorithmica*, 30, 312–333, 2001.

Schylberg, L., Computational Methods for Generalization of Cartographic Data in Raster Environment, doctoral thesis, Royal Institute of Technology, Stockholm, Sweden. 1993.

Su, B., Morphological Transformations for Generalization of Spatial Data Sets in Raster Format, PhD thesis, Curtin University of Technology, Perth, Australia, 1997.

Su, B. and Li, Z. L., An algebraic basis for digital generalization of area-patches based on morphological techniques, *Cartographic Journal*, 32(5), 148–153, 1995.

Su, B., Li, Z. L., Lodwick, G., and Müller, J. C., Algebraic models for the aggregation of area features based upon morphological operators, *International Journal of Geographical Information Science,* 11(3), 233–246, 1997.

Tsai, V. J. D., Delaunay triangulations in TIN creation: an overview and a linear-time algorithm, *International Journal of Geographical Information Systems*, 7(6), 501–524, 1993.

11 Algorithms for Displacement of Features

11.1 DISPLACEMENT OF FEATURES: AN OVERVIEW

Spatial conflicts are caused either by the nature of scale reduction or by the use of inadequate operators or algorithms in multi-scale transformations. In general, they can be classified into three categories, as shown in Figure 11.1.

These spatial conflicts need to be resolved when they occur. It is even better if they can be avoided. Those caused by the nature of scale reduction cannot be avoided and thus need to be resolved. However, those caused by inadequate algorithms should be avoided by using adequate algorithms. It can be noted that the "fighting" type of spatial conflicts is mainly caused by the use of inadequate operators and should be avoided. Therefore, a discussion of fighting conflicts is not necessary and is thus excluded from this text. The resolution of the "touching" and "coalescing" types of spatial conflicts is the main theme of this chapter.

To resolve these two types of spatial conflicts, an operation called displacement needs to be applied. Displacement can be classified into two categories: translation and modification. Figure 11.2 illustrates these two types of displacement. Figure 11.2a is the original representation (i.e., a map); Figure 11.2b shows the coalescence between the two features after scale reduction. This problem can be resolved either by an upward translation of the rectangular object as shown in Figure 11.2c or by a modification of the line feature as shown in Figure 11.2d. *Feature translation* has also been referred to as *feature transportation* by Rhind (1973).

Translation of a feature can be carried out in a single direction or in the normal directions of the boundaries of the features. The translation can be carried out for an individual feature or for all features. The translation of features in different directions and with different amounts of distance is referred to as *relocation* in this text and will be discussed in Section 11.4. Simple translation in a single direction with the same amount of distance is called *uniform translation* in this text and will be discussed in Section 11.2.

Modification can be carried out for linear features or for area features. For linear features, modification usually means a change in bends, while for area features it usually means a change in size. Modification can be carried out in a local context or globally for whole features. Modification of linear features will be discussed in Section 11.3.

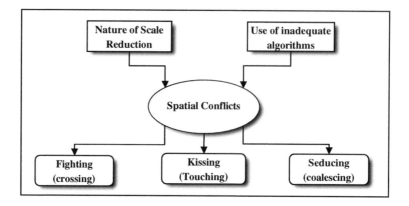

FIGURE 11.1 Three basic types of spatial conflicts (Reprinted from Li and Su, 1996. With permission.).

However, in this text, the change in size of area features is not discussed because it is a rather simple operation.

Most of the algorithms for relocation of features work with special cases. As one can imagine, there are too many special cases for discussion. Therefore, in this text, only generic algorithms are discussed.

Displacement is used to solve a conflict problem. Before displacement can be carried out, the potential conflicts should be detected, which, as pointed out by Monmonier (1987), can be ideally performed in a raster model. Su and Li (1997)

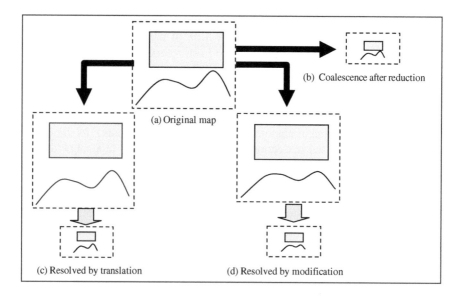

FIGURE 11.2 Two types of displacement: translation and modification.

have recommended the detection of conflicts by means of morphological operators, which work in raster mode. However, we do not include the detection of conflicts here because it is not an operation for multi-scale spatial representation.

11.2 ALGORITHMS FOR TRANSLATIONS OF FEATURES

11.2.1 UNIFORM TRANSLATION IN A SINGLE DIRECTION IN RASTER MODE

Li and Su (1996) have proposed the use of morphological operators for the uniform translation of features in a single direction. Figure 11.3 shows an example of using the dilation operator to translate the feature to the right direction. Li and Su also discovered that, in addition to dilation, erosion can be used for this purpose, and the translation to eight directions (i.e., right, left, up, down, upper right, upper left, lower left, lower right) is possible with a set of specially designed structuring elements. The model for translation is simply:

$$D = A \oplus T_d \tag{11.1}$$

or

$$E = A \ominus T_e \tag{11.2}$$

where A is an original feature and T_d and T_e are the two sets of the specially designed structuring elements as shown in Figure 11.4 and Figure 11.5.

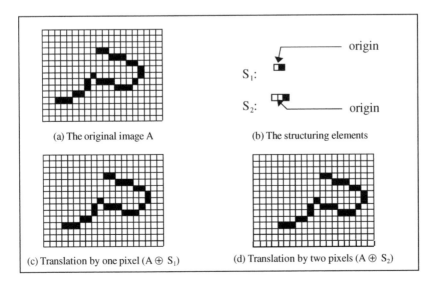

(a) The original image A

(b) The structuring elements

(c) Translation by one pixel (A $\oplus$ S$_1$)

(d) Translation by two pixels (A $\oplus$ S$_2$)

FIGURE 11.3 Feature translation to the right (Reprinted from Li and Su, 1996. With permission.).

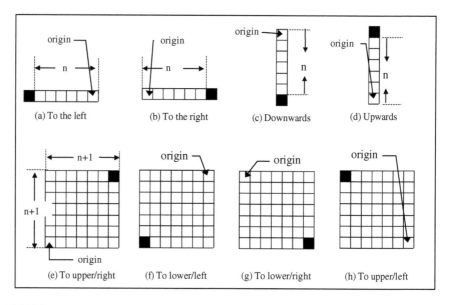

FIGURE 11.4 Structuring elements for translating objects using dilation, by n pixels in eight basic directions (Reprinted from Li and Su, 1996. With permission.).

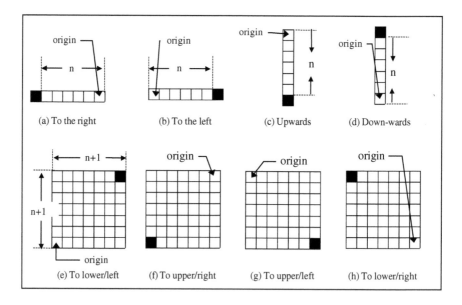

FIGURE 11.5 Structuring elements for translating objects using erosion, by n pixels in eight basic directions (Reprinted from Li and Su, 1996. With permission.).

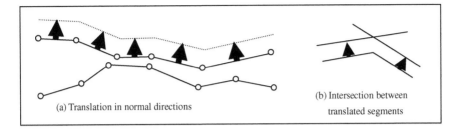

FIGURE 11.6 Displacement of features in normal directions.

This set of structuring elements shown in Figure 11.5 is exactly the same as the set used for dilation shown in Figure 11.4 except that the directions of movement are opposite. For example, in the case of dilation, the structuring element shown in Figure 11.4a is used for translating features to the left while, in the case of erosion, it is used for translating features to the right.

11.2.2 TRANSLATION IN NORMAL DIRECTIONS IN VECTOR MODE

In vector mode a feature is displaced piecewise in the corresponding normal directions. The translated pieces are joined by the intersecting points. Figure 11.6 illustrate the displacement in vector mode. This kind of operation is simple and is not worth further elaboration. Monmonier (1987) believed that "a truly efficient displacement model would require both vector and raster representation of some sort."

11.3 DISPLACEMENT BY PARTIAL MODIFICATION OF A CURVED LINE

11.3.1 PARTIAL MODIFICATION WITH A VECTOR BACKBONE

In raster mode the displacement moves pixels individually. Monmonier (1987) noticed that if special care is not taken, the so-called cascade effect can be created, which can cause a feature to lose its characteristics. He suggests the use of a vector "backbone of sorts but with controlled flexibility between adjoining vertebrae." Figure 11.7 shows this case. In this way, instead of merely taking a minor detour around a cell of overlap, a larger portion of a somewhat stiffened feature needs to be relocated.

11.3.2 PARTIAL MODIFICATION WITH MORPHOLOGICAL ALGORITHMS

Li and Su (1996) have proposed an algorithm for the modification of linear features, which can produce the backbone effect automatically. The basic idea is illustrated

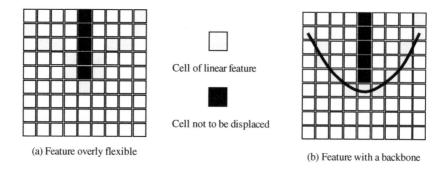

(a) Feature overly flexible

Cell of linear feature

Cell not to be displaced

(b) Feature with a backbone

FIGURE 11.7 Modification for displacement with a vector backbone.

in Figure 11.8. The concerned parts of both lines are dilated so as to become two overlapping strips. If one line is modified, then the overlapped part is cut from the corresponding strip. Then, the skeleton of this strip is derived and used to replace the original line.

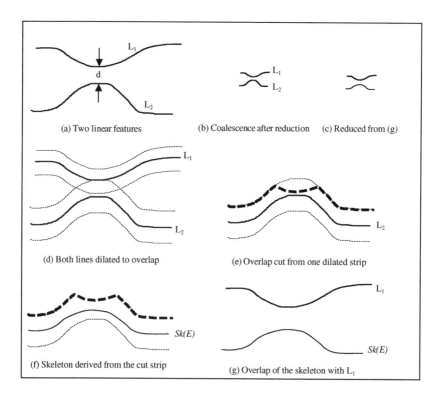

(a) Two linear features

(b) Coalescence after reduction   (c) Reduced from (g)

(d) Both lines dilated to overlap

(e) Overlap cut from one dilated strip

(f) Skeleton derived from the cut strip

(g) Overlap of the skeleton with L_1

FIGURE 11.8 Principle of partial modification of a curved line (Reprinted from Li and Su, 1996. With permission.).

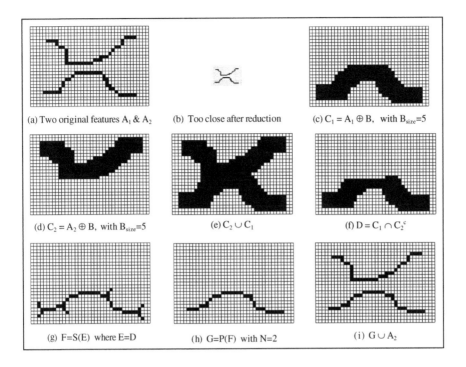

(a) Two original features A_1 & A_2 (b) Too close after reduction   (c) $C_1 = A_1 \oplus B$, with $B_{size}=5$

(d) $C_2 = A_2 \oplus B$, with $B_{size}=5$ (e) $C_2 \cup C_1$ (f) $D = C_1 \cap C_2^c$

(g) $F = S(E)$ where $E = D$ (h) $G = P(F)$ with $N = 2$ (i) $G \cup A_2$

FIGURE 11.9 Modification of one linear feature using morphological operators (Reprinted from Li and Su, 1996. With permission.).

The whole process of the *Li–Su modification algorithm* is illustrated in Figure 11.9, and the algorithm is as follows (Li and Su, 1996):

1. Dilate both linear features to obtain two area features, called strips in this text:

$$S_1 = L_1 \oplus B \tag{11.3}$$

$$S_2 = L_2 \oplus B \tag{11.4}$$

2. Cut off the overlapping area from the area strip dilated from the line feature to be modified:

$$D = S_1 \cap S_2^c \tag{11.5}$$

3. Obtain the skeleton of the area feature that is cut:

$$E = D \ominus B_1 \quad \text{(optional)} \tag{11.6}$$

$$F = Sk(E) \tag{11.7}$$

4. Perform a pruning operation to cut off parasitic branches:

$$G = P(F) \tag{11.8}$$

5. Obtain final result with an overlay operation:

$$R = G \cup L_2 \tag{11.9}$$

where L_1 and L_2 are two original line features.

The width of the strip (i.e., R) to be dilated is dependent on the current minimum separation between these two features (i.e., d in Figure 11.8a) and the desirable value of increase (e.g., Δd). The formula is as follows:

$$2R = (2 \times \Delta d + d) \tag{11.10}$$

If expressed in terms of the number of pixels, the size of the structuring element B is computed as follows:

$$B = INT \left(\frac{INT(2*R/\text{Pixel\_size} + 0.5)}{2} \right) \times 2 + 1 \tag{11.11}$$

where INT means to take the integer part of a value.

Figure 11.10 is a practical example of partial modification (Li and Su, 1997). Figure 11.10a is an Australian map at 1:20,000. Figure 11.10b shows two roads to be considered. If these two lines are thickened for reduction to scale 1:100,000, they will appear to be overlapping (Figure 11.10b,c). Therefore, the bend is modified using this algorithm, and the result is shown in Figure 11.10e.

The process of modification of both lines is similar. The only difference is that one needs to apply the modification process twice, once for each feature. However, in the process of modifying the second line (e.g., line L_2 in Figure 11.8), the width of the strips to be dilated becomes:

$$2R = (d + 2 \times \Delta d_2 + \Delta d_1) \tag{11.12}$$

where R is the width of the strip to be dilated, d is the current minimum separation between these two features (i.e., d in Figure 11.7a), and Δd_1 and Δd_2 are the desirable values of increase from modifying L_1 and L_2, respectively.

11.3.3 PARTIAL MODIFICATION BASED ON SNAKES TECHNIQUES

In Section 6.4.3 in Chapter 6 the application of snakes techniques for smoothing a line was described. In smoothing, the position of the snakes is represented by a parametric function expressed in Equation 6.15, and the internal energy expressed

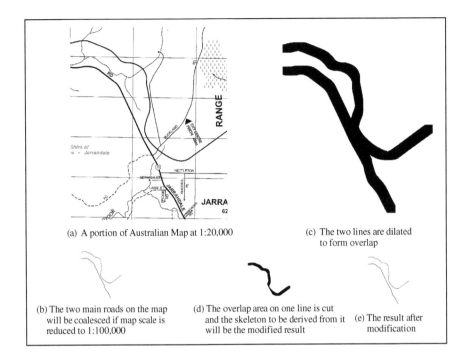

(a) A portion of Australian Map at 1:20,000

(c) The two lines are dilated to form overlap

(b) The two main roads on the map will be coalesced if map scale is reduced to 1:100,000

(d) The overlap area on one line is cut and the skeleton to be derived from it will be the modified result

(e) The result after modification

FIGURE 11.10 Partial modification of one line feature only; a practical example (Reprinted from Li and Su, 1996. With permission.).

by Equation 6.14 is then minimized. In the case of modification for displacement, the difference between the initial and the resultant (displaced) line is employed (Burghardt and Meier, 1997; Burghardt, 2005). To contrast the difference between smoothing and displacement, these two functions are put together as follows:

$$g_s(l) = \begin{cases} x(l) \\ y(l) \end{cases} \tag{11.13}$$

$$g_d(l) = \begin{cases} x(l) - x_0(l) \\ y(l) - y_0(l) \end{cases} \tag{11.14}$$

where $g_s(l)$ is the function used for smoothing, $g_d(l)$ is the function used for displacement, $x_0(l)$ and $y_0(l)$ represent the coordinates of the original line, and $x(l)$ and $y(l)$ represent the coordinates of the displaced or smoothed line.

Bader and Barrault (2000) introduced a "user-defined displacement which is imposed as boundary conditions to one of the snake" (Figure 11.11). In this way, partial modification of a line to a desirable amount is under control, which is

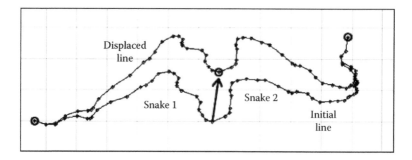

FIGURE 11.11 Partial modification by snakes with user-defined displacement (Reprinted from Bader and Barrault, 2000. With permission.).

similar to the Li–Su modification algorithm (Li and Su, 1996). In Figure 11.11 the arrow indicates the vector displacement defined, and the two circles indicate anchor points that are also defined by the user and will become the tails of the two snakes. Both snakes have their tails set to zero and their heads set to the defined displacement.

At the same time, Bader and Barrault (2000) also noticed the problems with this modified work as follows:

1. The result is dependent on the position of the tail.
2. The characteristics of the line are not taken into consideration.
3. There is no guarantee of avoiding self-intersection, and it is not possible to propagate the displacement over a junction.

To overcome these drawbacks, Bader and Barrault (2000) added a term called *attraction* to the model to produce an extended snakes model as follows:

$$E(g) = \int_l \left(\frac{1}{2} \left(\psi |g(s)|^2 + \alpha |g'(s)|^2 + \beta |g''(s)|^2 \right) \right) ds \qquad (11.15)$$

where ψ penalizes the line, as long as it does not coincide with the original position, to make the line strive to fall back to its initial position (Bader and Barrault, 2000). The relationship among these three terms is as follows:

$$\psi \times g(s) + \alpha \times g'(s) + \beta \times g''(s) = 0 \qquad (11.16)$$

Bader and Barrault also suggested the adjustment of α and β instead of ψ for different parts of a line. High α and β values can be use for the protection of the

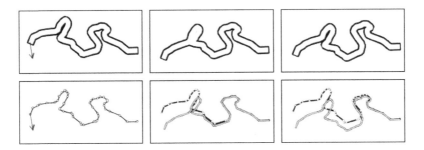

FIGURE 11.12 Improved propagation of displacement in a line (right) to avoid an unacceptable result (middle). (Reprinted from Bader and Barrault, 2000. With permission.).

deformation of certain (parts of) curves, and low α and β values can be used for those parts of a line that are oriented in the displacement direction. Figure 11.12 shows an example of the improved snakes model.

11.4 ALGORITHMS AND MODELS FOR RELOCATION OF FEATURES

This section only outlines the basic principles. The detailed mathematical deductions are omitted because they are not more models than algorithms in a practical sense.

11.4.1 RELOCATION OF FEATURES WITH DISPLACEMENT FIELDS

Ai (2004) proposed the concept of *displacement fields* for the relocation of buildings. This field is analogous to the electromagnetic field (or mass field) in physics. That is, a field (or force) originates from one or more features that are to be kept unmoved. In this method it is assumed that every feature receives a driving force to move itself from its original position and at the same time exerts a force with a reduced magnitude to push the neighboring features to move. To make this method work, one needs to define:

1. The source of the force
2. The propagation behavior (direction and decay function)
3. The associated neighbors that will receive the propagated force

The neighboring relation is described by the adjacency of the Voronoi regions (Figure 11.13b). The distance between Vooronoi neighbors is defined by the order of adjacency, that is, the Voronoi-based K-order relations developed by Chen et al. (2004). The K-order is then represented in the form of isolines (Figures 11.13c and 11.13d). In this way the features located on the same isoline receive a force of identical magnitude. The isolines closer to the source of force represent larger magnitudes in force.

(a) A portion of Shenzhen City map

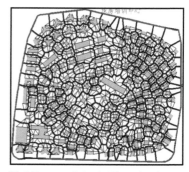

(b) Adjacency relation by Voronoi regions

(c) Isolines of equal K-order neighbours from A

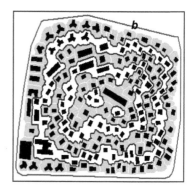

(d) Isolines of equal order from boundary b

FIGURE 11.13 Building relocation by a displacement field (Reprinted from Ai, 2004. With permission.).

The movement direction of each building from the source of force that is caused by the propagation force can be computed by the additional operation of vectors along the normal direction of Voronoi edges. The amount of displacement is determined as follows:

$$d = f(k) = d_0 - a \times k \tag{11.17}$$

where d is the displacement magnitude, d_0 is the maximum amount of displacement, k is the order of adjacency, and a is a constant. This is a linear function, and other nonlinear functions may also be used.

Figure 11.14 (See color insert following page 116) shows a practical example of the relocation of buildings shown in Figure 11.13a. The relocation is caused by street widening, which is used by Ai (2004). In this case, the source of force is the street boundary. The force is propagated from the street as a boundary to the inner buildings. The magnitude of displacement gradually becomes smaller toward the center of the block. Figure 11.14a shows the magnitude and direction of movement, and Figure 11.14b shows the final result after relocation.

It is possible for buildings to become coalesced or overlapped after relocation. In this case, the local repulsive force originating from close neighbors should also be considered.

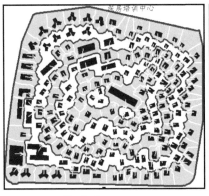 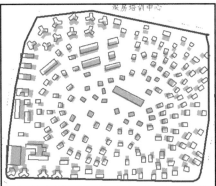

(a) The displacement vectors   (b) The result after relocation

FIGURE 11.14 (See color insert following page 116) Displacement field applied to building relocation caused by street widening (Reprinted from Ai, 2004. With permission.).

11.4.2 RELOCATION OF FEATURES WITH FINITE ELEMENTS

Finite Element analysis was introduced originally as a method for solving structural mechanical problems, which was later recognized as a general procedure for numerical approximation to all physical problems that can be modeled by a differential equation description. Finite Element analysis has also been applied to the description of physical form changes in biologic structures particularly in the area of growth and development and restorative dentistry. (Vandana and Kartik, 2005)

In spatial information science, it has also been used for interpolation in digital terrain modeling in the 1980s (Ebner et al., 1980) and for feature relocation by Hojholt (2000).

The *finite element method* (FEM) has been widely used to solve the partial differential equations of complex structure. *Finite element* means that the study domain is discretized into a number of geometric elements that are connected via nodes. The element can be grid cells or triangular cells. Each element is given a specific internal strain function. Using these functions and the geometry of the element, the equilibrium equations between the external forces acting on the element and the displacements occurring on its nodes can be determined (Vandana and Kartik, 2005). In this way a complex problem is redefined as the summation of a series of interrelated simpler problems.

In the work by Hojholt (2000) triangular cells are used because map features can easily be interconnected by Delaunay triangulation (e.g., Jones et al., 1995). Each triangle can be moved and deformed. The movement includes translation and rotation. The deformation of a triangle is completely described by the movement of the three vertices.

The movement and deformation of a feature is caused by a force. A feature can be allocated a force to expand itself and to cause other surrounding features to move and to deform. The deformation and movement are defined by vectors in the X and Y directions. However, all features are allocated a certain degree of stiffness and some boundary conditions. It is also possible to define anchor points that are not to be moved during operation. In this way, a balance is created. This is the basic idea of the finite element method. More detailed mathematical deduction lies outside the scope of this text and can be found in Hojholt (2000) and Vandana and Kartik (2005).

11.4.3 RELOCATION OF FEATURES WITH LEAST-SQUARES ADJUSTMENT

Least squares adjustment is a well-known technique used to obtain a solution for a set of overdetermined equation systems. It has been applied to map generalization by Harrie (1999), Sarjakoski and Kilpelainen (1999), and Sester (1999).

Harrie (1999) claims that in his work some ideas were borrowed from the snakes and finite element methods. He identified five types of displacement behavior:

1. Movable-point features: Point features that are allowed to move.
2. Immovable-point features: Point features that are not allowed to move.
3. Immovable line and area features: Line and area features that are not allowed to change at all.
4. Stiff line and area features: Line and area features that are not allowed to move but are not distorted.
5. Ordinary line and area features: Line and area features that are allowed to move and distort to a small degree.

Different weights are assigned to different types of features. A set of five types of constraints is also used:

1. Movement constraints: Feature movement as little as possible.
2. Stiffness constraints: Internal geometry invariant.
3. Curvature constraints: Curvature of a line or boundary of an area that is not changed.
4. Crossing constraints: Intersecting angle of two lines that are not changed.
5. Spatial conflict constraints: No spatial conflicts allowed.

In the work by Sester (2000), the constraints are:

1. Form parameters: Side, angles and orientation.
2. Distance between objects: Minimum separation between features with zero for the features to merge.
3. Additional parameters: Coordinates.

All these constraints can be used to form conditional equations. In the end, the sum of the squared displacements is minimized to achieve the overall objectives:

1. Solve all spatial conflicts.
2. Preserve the characteristics of the spatial distribution of the features.

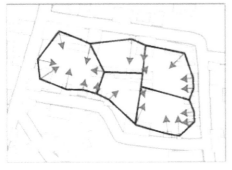
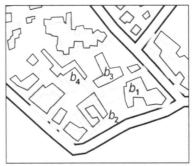

(a) Ductile truss structure to tight buildings as a pattern   (b) Example of result after relocation

FIGURE 11.15 Building relocation over a ductile truss (Reprinted from Bader et al., 2005. With permission.).

11.4.4 RELOCATION OF FEATURES WITH A DUCTILE TRUSS AND FINITE ELEMENTS

Bader et al. (2005) noted that earlier methods directly displace the individual buildings and then proposed a new method for building relocation. In their method a *ductile truss* is superimposed onto the group of buildings to be relocated (see Figure 11.15a). The ductile truss is built upon the minimum spanning tree (MST) of the building centroids, which establishes the topological relationship between the buildings. Onto the MST, other important spatial relations are then added to enrich the data structure. This truss structure is designed to preserve the spatial relations between buildings. The truss is assumed to consist of elastic beams (i.e., the spans in the MST), which can be modeled by the finite elements method. A degree of elasticity is assigned to each beam based on the separations between buildings and their parallelism. Then the truss is deformed to minimize the energy until the user-defined distance is achieved. Forces for the deformation of the truss come from the boundary of the area (e.g., roads) and the coalescence of buildings. Figure 11.15b is an example of building relocations by this method.

REFERENCES

Ai, T. H., A displacement of building cluster based on field analysis, *Acta Geodaetica et Cartographica Sinica*, 33(1), 89–94, 2004.

Bader, M. and Barrault, M., Improving snakes for linear feature displacement in cartographic generalization, GeoComputation 2000. http://www.geocomputation.org/2000/GC034/ Gc034.htm.

Bader, M., Barrault, M., and Weibel, R., Building displacement over a ductile truss, *International Journal of Geographical Information Science*, 19(8-9), 915–936, 2005.

Burghardt, D., Controlled line smoothing by snakes, *GeoInformatica*, 9(3), 237–252, 2005.

Burghardt, D. and Meier, S., Cartographic displacement using the snakes concept, in *Semantic Modeling for the Acquisition of Topographic Information from Images and Maps*, Foerstner, W. and Pluemer, L., Eds., Birkhaeuser-Verlag, Basel, Switzerland, 1997, pp. 59–71.

Chen, J., Zhao, R. L., and Li, Z. L., Voronoi-based K-order neighbour relations for spatial analysis, *ISPRS Journal of Photogrammetry and Remote Sensing*, 59 (1–2), 60–72, 2004.

Ebner, H., Hofmann-Wellenhof, B., Reiss, P., and Steidler, F., HIFI: a minicomputer program package for height interpolation by finite elements, *International Archives of Photogrammetry and Remote Sensing*, 23(IV), 202–241, 1980.

Harrie, L. and Sarjakoski, T., Simultaneous graphic generalization of vector data sets, *GeoInformatica*, 6(3), 233–262, 2002.

Harrie, L. E., The constraint method for solving spatial conflicts in cartographic generalization, *Cartography and Geographic Information Systems*, 26(1), 55–69, 1999.

Hojholt, P., Solving space conflicts in a map generalization: using a finite element method, *Cartography and Geographic Information Science*, 27(1), 65–73, 2000.

Jones, C. B., Bundy, G. L., and Ware, J. M., Map generalization with a triangulated data structure, *Cartography and Geographic Information Systems*, 22(4), 317–331, 1995.

Li, Z. L. and Su, B., Algebraic models for feature displacement in the generalization of digital map data using morphological techniques, *Cartographica*, 32(3), 39–56, 1996.

Li, Z. L. and Su, B., Some basic mathematical models for feature displacement in digital map generalization, in *Proceedings of ICC'97*, Vol. 1, June 23–27, 1997, Stockholm, pp. 452–459.

Monmonier, M., Displacement in vector- and raster-mode graphics, *Cartographica*, 24(4), 25–36, 1987.

Nickson, B., Automated cartographic generalization for line features, *Cartographica*, 25(3), 15–66, 1988.

Rhind, D., Generalization and realism with automated cartographic system, *Canadian Cartographer*, 10(1), 51–62, 1973.

Sarjakoski, T. and Kilpelainen, T., Holistic cartographic generalization by least squares adjustment for large data sets, in *Proceedings of 19th International Cartographic Conference*, August 16–20, 1999, Ottawa, Canada, 1999, pp. 1091–1098.

Sester, M., Generalization based on least squares adjustment, in *Proceedings of the International Cartographic Conference*, ICA, Ottawa, 1999. (CD-Rom)

Sester, M., Generalization based on least squares adjustment, *International Archives of Photogrammetry and Remote Sensing*, XXXIII(Part B4), 931–938, 2000.

Su, B. and Li, Z. L., Morphological transformation for detecting spatial conflicts in digital generalization, in *Proceedings of ICC'97*, Vol. 1, June 23–27, 1997, Stockholm, 1997, pp. 460–468.

Vandana, K. L. and Kartik, M., Finite Element Method: Perio-endo Concept, http://medind.nic.in/eaa/t04/i2/eaat04i2p38.pdf, 2005.

12 Algorithms for Transformations of Three-Dimensional Surfaces and Features

12.1 ALGORITHMS FOR TRANSFORMATIONS OF THREE-DIMENSIONAL FEATURES: AN OVERVIEW

Traditional spatial representations are in two-dimensional (2-D) form and static mode. However, dynamic and multidimensional representations are now being more widely used.

Terrain surfaces are popularly represented in digital and 3-D forms, leading to terms such as digital terrain models (DTMs) and digital elevation models (DEMs). The DEM is normally used to refer to the digital representation of elevation values in grid form. DTM is a concept covering other information in addition to elevation in grid form. Triangular irregular network (TIN) is another form widely used in practice. DTMs, together with digital contour data, form a major part of a national spatial data infrastructure. A section in this chapter is devoted to the multi-scale representation of DTM surfaces.

The features on the terrain surface have also been popularly represented in 3-D form for various applications such as virtual landscapes, virtual battle fields, environmental analysis, and geographical analysis. The 3-D representation of buildings to form digital city models is a hot topic. In this chapter, another section is devoted to the multi-scale representation of 3-D buildings.

12.2 ALGORITHMS FOR TRANSFORMATIONS OF DTM SURFACES

12.2.1 MULTI-SCALE TRANSFORMATION OF DTM SURFACES: AN OVERVIEW

For 3-D surfaces represented by a DTMs, two types of multi-scale representation have been differentiated (Li et al., 2005): metric and visual. The first one is map-like, emphasizing the metric quality, and thus, is used for measurement. The *metric multi-scale representation* of the DTM means to automatically derive DTM data suitable for any smaller scale representation from the DTM at the largest scale,

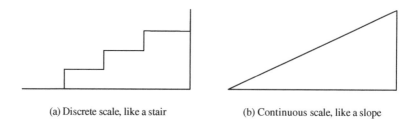

(a) Discrete scale, like a stair   (b) Continuous scale, like a slope

FIGURE 12.1 Steps and linear slope compared to discrete and continuous transformations.

which is updated continuously. Such a process is called *generalization* and it is applied uniformly across the whole area covered by the DTM data so that all data points within the area have a uniform accuracy. The other type, for visual impressions only (e.g., for computer graphics and games), is called *visual multi-scale representation*. On the same representation, the scale is not the same over the whole area and is a function of viewing distance. In other words, the level of detail (LoD) on a representation varies from place to place. This kind of approach is simply called LoD in computer graphics. In some of the literature, this is also called *view-dependent LoD* and, by contrast, metric multi-scale representation is called *view-independent LoD*.

There are also two types of transformations in scale: discrete and continuous. *Discrete* means there are only a few scales available, for example, 1:10,000, 1:100,000, and 1:1,000,000. The transformation jumps from 1:10,000 to 1:100,000, then to 1:1,000,000. Discrete transformation is like fixed steps in a staircase (Figure 12.1a). *Continuous* means the transformation can be to any scale, for example, 1:50,000 or 1:56,999, although in practice some scales are not used (e.g., 1:56,999). Continuous transformation is like a linear slope (Figure 12.1b).

The approaches to multi-scale representation of digital terrain surfaces are summarized in Figure 12.2 (Li et al., 2005).

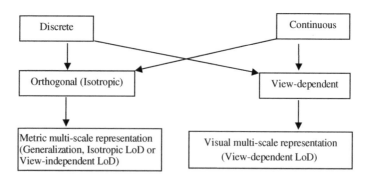

FIGURE 12.2 Alternative approaches for multi-scale representation of DTM data (Reprinted from Li, et al., 2005).

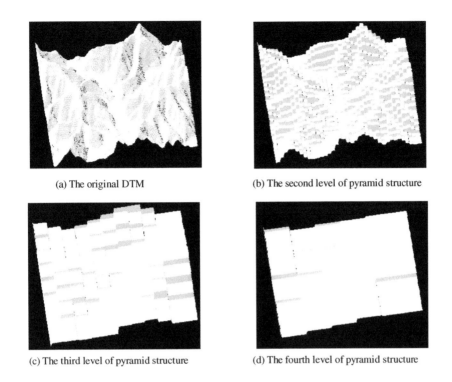

(a) The original DTM   (b) The second level of pyramid structure

(c) The third level of pyramid structure   (d) The fourth level of pyramid structure

FIGURE 12.3 Multi-scale representation of a DTM surface at discrete scales.

Since the late 1970s, metric multi-scale representation of DTM surfaces has been a research topic. Weibel (1987) proposed six criteria for the evaluation of methodologies:

- To run as automatically as possible
- To perform a broad range of scale changes
- To be adaptable to the given relief characters
- To work directly on the basis of the DTM
- To enable an analysis of the results
- To provide the opportunity for feature displacement based on the recognition of the major topographic features and individual landforms (major scale reduction)

If the set of criteria proposed by Weibel (1987) is used to evaluate a multi-scale representation at the discrete scale (de Floriani, 1989; de Berg and Dobrindt, 1998), for example, pyramidal representations as shown in Figure 12.3, the results are not very good. The most serious shortcoming is that it only produces a fixed number of scales, at least for hierarchical grid networks, that is, 2x, 4x, 8x, 16x, ..., scale reduction. This approach is only convenient for data structures with which visualization of DTM surfaces could be sped up, but it is not a generic approach for multi-scale representation. Therefore, this approach is omitted in this text.

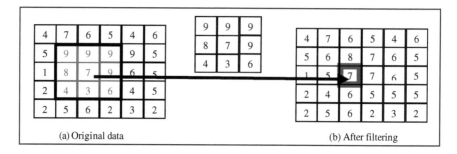

(a) Original data   (b) After filtering

FIGURE 12.4 Low-pass filtfor smoothing a DTM surface.

In metric multi-scale representation at a continuous scale, some methods have been in use, e.g., filtering (e.g., Loon, 1978); the simplification of structure lines (e.g., Wu, 1997), an adaptive smoothing (based on a combination of these two) (Weibel, 1987, 1992), and scale-driven generalization based on the Natural Principle (Li and Li, 1999). In this text, transformations accomplished through the simplification of structure lines are not discussed because they are at more of an approach level than an algorithmic level.

12.2.2 METRIC MULTI-SCALE REPRESENTATION THROUGH FILTERING AND RESAMPLING

The first paper on numerical generalization by Tobeler (1966) was about the application of a low-pass filer for smoothing a set of digital data, which could be elevation or any other 3-D variations (e.g., temperature). Simple and weighted averaging can be used, and the result of a simple averaging is shown in Figure 12.4. The mathematical formula for simple averaging is as follows:

$$z = \frac{\sum_{i=1}^{N} z_i}{M} \tag{12.1}$$

where N is the total number of reference points used for the averaging operation and z_i ($i \in 1, N$) is the height of the ith reference point. For example, if the points in a 3×3 window centered at P are selected for averaging, as reference points for interpolation of point, and their heights are 9, 9, 9, 8, 7, 9, 4, 3, and 6, then the average is 7. In the filtering processing, a template (e.g., a 3×3 window) is moved over the DTM cells one by one in row and column directions, and the averaging of the heights within the template is used to represent the height value at this position in the output DTM.

In simple averaging, no matter how close a reference is to the interpolation point, the weight is still the same as that of the others. This equal weighting seems unfair to reference points that are closer to the central point. This leads to a *weighted averaging* as follows:

$$z = \frac{\sum_{i=1}^{N} w_i z_i}{\sum_{i=1}^{N} w_i} \tag{12.2}$$

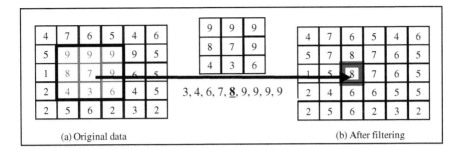

FIGURE 12.5 A median filter for smoothing a DTM surface.

where w_i is the weight for the ith point, which may or may not be different from the others'. Normally, the weight is a function of distance, for example, inverse proportion. If the size of the template (window) is big, the polynomial surface may also be fit onto the cells in the template to obtain a value for the central position of the template.

The principle of averaging used in this section is identical to the averaging used for line smoothing discussed in Chapter 6, except that a 2-D window is used in former case and a 1-D window in the latter case.

A *median filter* can also be used for smoothing. The principle of a median filter is illustrated in Figure 12.5. The operational process is the same as that for the low-pass filter except that the median of the values in the template is used in this case (instead of the average).

In the case of low-pass and median filters, the cell size (or grid interval) is not changed. In practice, a smoothing effect can also be produced by changing the cell size. *Resampling* is one of these techniques. Figure 12.6 shows an example. In this figure the cell size of the new DTM grid is 1.5 times the original. Two interpolation techniques have been used: nearest neighbor and bilinear. If the cell size of the new DTM grid is four (or more) times larger than the original, bilinear interpolation may also be used.

Another technique is called *aggregation*. With aggregation, all cells in an $n \times n$ window are aggregated into a new cell. The value of the new cell could take the mode, median, or average of the cells within the window. It is also possible to select

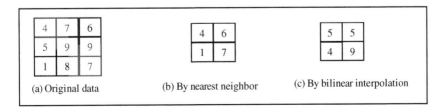

FIGURE 12.6 Resampling for smoothing a DTM surface.

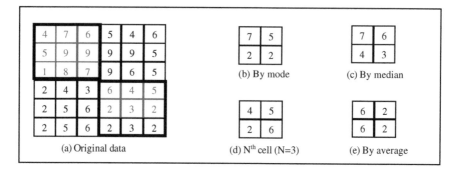

(a) Original data

(b) By mode

(c) By median

(d) N^{th} cell (N=3)

(e) By average

FIGURE 12.7 Aggregation for smoothing a DTM surface.

only the pixels at the nth row and column from the original DTM grid. Figure 12.7 shows these results.

12.2.3 METRIC MULTI-SCALE REPRESENTATION BASED ON THE NATURAL PRINCIPLE

The natural principle developed by Li and Openshaw (1993) has been described in Chapter 3. This principle was successfully implemented for digital map generalization (Li and Openshaw, 1992).

The process of applying the natural principle to metric multi-scale representations of DTMs is illustrated in detail in Figure 3.5 in Chapter 3. In that diagram the scale of the height of the image plane is computed from the scales of input and output DTMs. The area to be ignored each time is determined by the projection of the rays from the image plane. It can be approximated by a convolution process, carried out cell by cell on the input DTM. Each time a template with a size equal to the smallest visible size (SVS) is placed onto a cell of the input DTM, all the cells within this template will be used to estimate the height of this cell in the output DTM. The computation process is illustrated in Figure 12.8. Li and Li (1999) used

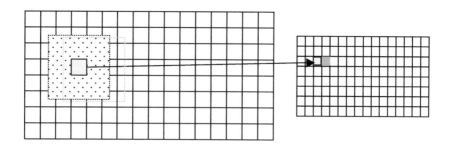

FIGURE 12.8 The movement of an SVS template over the input DTM, pixel by pixel, within which all spatial variations are ignored; for example, 9 cells (16 nodes) are aggregated into 1 in this example.

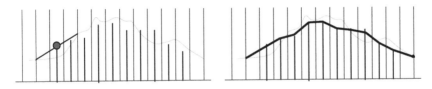

FIGURE 12.9 Aggregation of pixels in an SVS template (4 × 4 pixels). (The average of the boundary pixels only is used in this example.)

two methods. One is to take the average of all cells within the template, and the other is to take the average of the cells along the edges of the template.

Figure 12.9 shows the generalization of a profile from a DTM surface based on the method described in this section. The 1:20,000 scale DTM as shown in Figure 12.8 was generalized to produce DTMs at 1:50,000, 1:100,000, and 1:200,000 scales. It is clear that the surface becomes smoother as the scale becomes smaller. The contour plots of these DTM surfaces are given in Figure 12.10. It is clear that contour lines are smoother when the scale is smaller (Figure 12.11). This is in accordance with the natural generalization.

It must be noted here that this method is similar to low-pass filtering, but they differ in many aspects. First, the template can be moved gradually pixel by pixel or jumped from one pixel to another by a distance smaller than the template size.

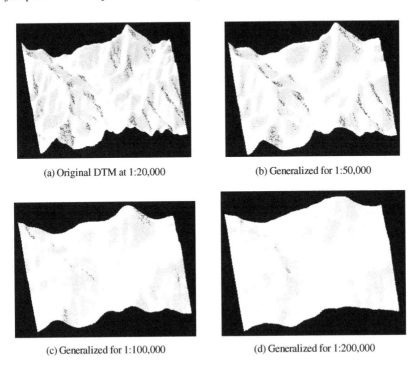

(a) Original DTM at 1:20,000   (b) Generalized for 1:50,000

(c) Generalized for 1:100,000   (d) Generalized for 1:200,000

FIGURE 12.10 Transformation of a DTM surface from 1:10,000 to 1:200,000, based on the natural principle.

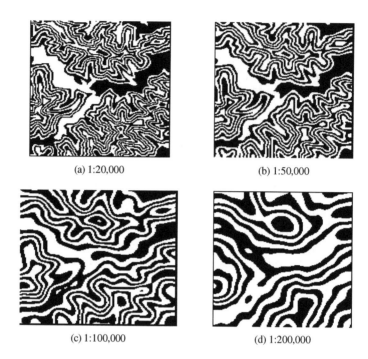

(a) 1:20,000   (b) 1:50,000

(c) 1:100,000   (d) 1:200,000

FIGURE 12.11 Contour representation of DTM surfaces at different scales.

In an extreme case there may be no overlap between templates, leading to the simple pyramidal structure. Second, the size of the template is computed from the values of the source and target scales. Third, the SVS template should be projected onto the DTM surface, and the process by a convolution is only an approximation. Fourth, the template may take a different value, for example the average of boundary pixels.

In this method the degree of generalization is the only concern; the resulting grid size of the final DTM is not a concern. If desirable, a quad-tree structure may be used to represent (or compress) the generalized DTM.

This method is equally applicable to triangular irregular networks (TINs). In such a process, the height of each triangular vertex is modified. The new height value is the average of the heights of all the vertices within the corresponding SVS or the average of the intersecting points between the TIN surface and the projected SVS, as shown in Figure 12.12.

12.2.4 VISUAL MULTI-SCALE REPRESENTATION THROUGH VIEW-DEPENDENT LoD

As discussed previously, the LoD technique is used for computer graphics and DTM visualization. The basic idea of LoD is simple. More detail is used for scenes and objects closer to the viewpoint, and less detail is used for scenes and objects farther from the from viewpoint.

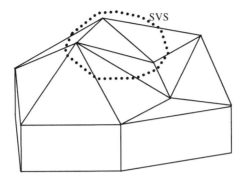

FIGURE 12.12 Generalization of a TIN surface based on the natural principle.

If the DTM is represented in a grid form, then coarser grid cells will be used for the representation of the scenes and objects that are more distant from the viewpoint. If the DTM is triangular, then coarser triangular cells will be used for the representation of the scenes and objects that are more distant from the viewpoint. Coarser cells, either grid or triangular, are generated by a number of operations such as *collapse* and *removal*. Figure 12.13 illustrates four of the basic operations for the simplification of a triangular network for LoD purposes:

Vertex removal: A vertex in the triangular network is removed and new triangles are formed (Figure 12.13a).

Triangle removal: A complete triangle with three vertices is removed and new triangles are formed (Figure 12.13b).

Edge collapse: An edge with two vertices is collapsed to a point and new triangles are formed (Figure 12.13c).

Triangle collapse: A complete triangle with three vertices is collapsed to a point and new triangles are formed (Figure 12.13d).

Now the question is, "When do we use these operations?" This concerns the constraints of simplification. Two constraints were used: (a) the number of very important points (VIPs) to be retained and (b) the allowable accuracy loss. These two constraints can also be used to simplify the DTM data for the generation of view-dependent LoD, leading to two distinct approaches. The one that uses allowable error as the constraint is also called *fidelity-based simplification,* and the one that uses the number of triangles as the constraint is also called *budget-based simplification* (Luebke et al., 2003).

One of the first real-time continuous LoD algorithms for terrain grids was the early work of Lindstrom et al. (1996) as cited by Luebke et al. (2003). The simplification scheme involves a vertex removal approach in which a pair of triangles is reduced to a single triangle. Figure 12.14 illustrates the principle. In this figure, ΔABC and ΔBCD are the two triangles considered. If vertex *C* is removed, then there will be a vertical error (δ) from *C* to line *A-D*. If δ on the screen is smaller

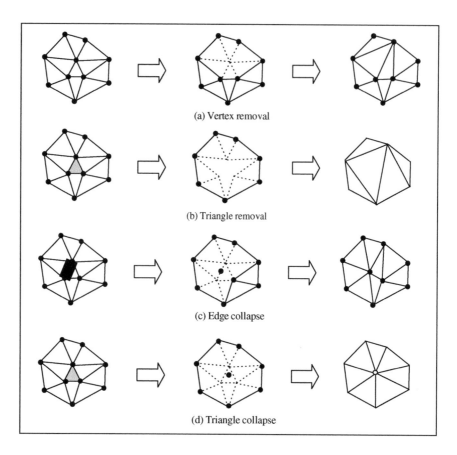

(a) Vertex removal

(b) Triangle removal

(c) Edge collapse

(d) Triangle collapse

FIGURE 12.13 Basic geometric operations for simplification of the triangular network.

than a threshold, then vertex C can be removed. This is the working principle of a fidelity-based simplification.

An extremely popular continuous LoD algorithm is the real-time optimally adapting meshes (ROAM) algorithm developed by Duchaineau et al. (1997). A continuous mesh is produced by applying a series of split and merge operations on a binary triangle tree (Figure 12.15). Again, screen-based geometry error is used as

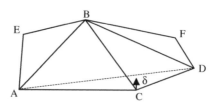

FIGURE 12.14 Fidelity-based LoD.

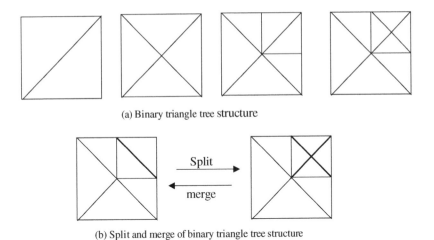

(a) Binary triangle tree structure

(b) Split and merge of binary triangle tree structure

FIGURE 12.15 Split-merge of a binary triangle tree in the ROAM algorithm.

a threshold for split and merge operations. Figure 12.16 is an example of the LoD of a terrain surface by this method.

The ROAM algorithm includes a number of other interesting features and optimizations, including an incremental mechanism used to build triangle strips. Real-time display of complex surfaces is provided by dynamically computing a multiresolution triangular mesh for each view. The meshes minimize geometric distortions on the screen while maintaining a fixed triangle count. Pop-ups are minimized in several ways, and efficient mesh corrections ensure that selected lines of sight or object proximity are correctly represented. An incremental priority-queue algorithm uses frame-to-frame coherence to quickly compute these optimal meshes (Duchaineau et al., 1997).

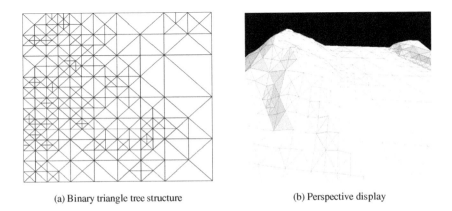

(a) Binary triangle tree structure

(b) Perspective display

FIGURE 12.16 LoD of a DTM by the ROAM algorithm (Reprinted from Li, et al., 2005).

12.3 ALGORITHMS FOR TRANSFORMATION OF 3-D FEATURES

Most research related to 3-D features has been done on buildings. The operations are still similar to those for 2-D features. Bunching and in-joining have been used by Bai and Chen (2001) in a way similar to the typification and aggregation operations for 2-D representations (Figure 12.17). However, some terms may have slightly different meanings. This section provides only a brief discussion of these operations.

12.3.1 TRANSFORMATION OF INDIVIDUAL BUILDINGS

Exaggeration is an operation that appears in the literature. Bai and Chen (2001) have used it to refer to the enlarged size of some characteristics of a building, for example, the doors and chimney of a building instead of the building itself. Such an exaggeration might be better defined as *thematic exaggeration* (Figure 12.17). Bai and Chen (2001) have also used exaggeration to refer to an increase in the size of some small features such as lamp posts along streets.

Displacement is another operation used for 3-D building features. Bai and Chen (2001) have used it to refer to the translation of a building to keep a minimum separation between the building and the road.

Simplification has been used to remove minor characteristics of a building (such as a door) and to eliminate small features (e.g., posts). The first might be better termed *thematic elimination. Omitting* has also been used to refer to the removal of small features (Thiemann, 2002). Other researchers have used *simplification* to refer to making the outline of the building simpler (Figure 12.18).

Forberg and Mayer (2002) have used *tapering* to refer to the flattening of inclined roof planes and then to subsequently removing the ridge line (Figure 12.19).

Forberg and Mayer (2002) have used the term *squaring* to refer the operation of making a not-rectangular building appear rectangular (Figure 12.20a). *Parallel shifting* is shifting the parallel facets of a building toward each other to make the small parallel facets into one single facet (Figure 12.20b).

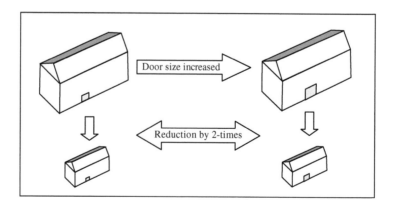

FIGURE 12.17 Thematic exaggeration: enlargement of the door of a building.

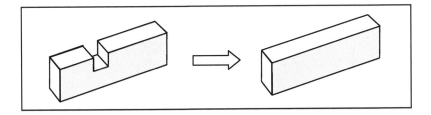

FIGURE 12.18 Simplification of the outline of a building.

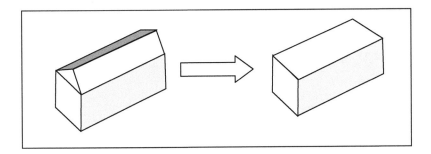

FIGURE 12.19 Tapering the roof of a building.

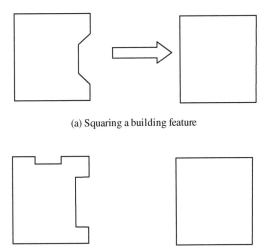

(a) Squaring a building feature

(b) Shifting of parallel facets

FIGURE 12.20 Squaring a building.

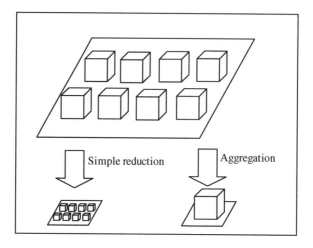

FIGURE 12.21 Aggregation of buildings.

12.3.2 Transformation of a Set of Buildings

Bai and Chen (2001) have used the term *bunching and in-joining* in a way similar to the typification and aggregation operations for 2-D representations (Figure 12.21). *Typification* has been defined by Thiemann (2002) as the replacement of *m* features by *n* features. It is identical to 2-D representation.

Forberg and Mayer (2002) have used the term *split* to refer to the separation of weakly connected components of a building and *merge* for the process of combining closely located buildings into one building (Figure 12.22). The merge is similar to the aggregation used by Thiemann (2002), who defines it as replacing *n* features with one feature.

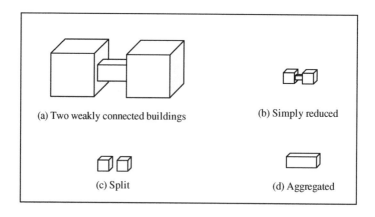

FIGURE 12.22 Split-and-merge of buildings.

REFERENCES

Bai, F. W. and Chen, X. Y., Generalization for 3D GIS, in *Proceedings of the 3rd ISPRS Workshop on Dynamic and Multi-dimensional GIS*, Chen, J., Chen, X. Y., Tao, C., and Zhou, Q. M., Eds., 2001, pp. 8–11.

de Berg, M. and Dobrindt, T. G., On the levels of detail in terrain, *Geographical Models and Image Processing*, 60(1), 1–12, 1998.

de Floriani, L., A pyramid data structure for triangle-based surface representation, *IEEE Computation Graphics and Applications*, 9, 67–68, 1989.

Duchaineau, M., Wolinsky, M., Sigeti, D. E., Miller, M. C., Aldrich, C., and Mineev-Weinstein, M. B., ROAMing terrain: real-time optimally adapting meshes, in *Proceedings of the 8th Conference on Visualization '97*, October 18–24, 1997, Phoenix, AZ, 1997, pp. 81–88.

Forberg, A. and Mayer, H., Generalization of 3D building data based on scale-space, *International Archives of Photogrammetry and Remote Sensing*, XXXIV(4), 225–230, 2002.

Li, Z. L. and Li, C., Objective representation of DTM surface in scale dimension, in *Proceedings of Joint ISPRS Commission Workshop on Dynamic and Multi-Dimensional GIS*, Beijing, China, 1999, pp. 17–22.

Li, Z. L. and Openshaw, S., Algorithms for automated line generalisation based on a natural principle of objective generalisation, *International Journal of Geographic Information Systems,* 6(5), 373–389, 1992.

Li, Z. L. and Openshaw, S., A natural principle for objective generalization of digital map data, *Cartography and Geographic Information Systems*, 20(1), 19–29, 1993.

Li, Z. L., Zhu, Q., and Gold, C., *Digital Terrain Modelling: Principles and Methodology*, CRC Press, Boca Raton, FL, 2005.

Lindstrom, P., Koller, D., Ribarsky, W., Hodges, L. F., Faust, N., and Turner, G., Real-time continuous level of detail rendering of height fields, in *Proceedings of SIGGRAPH 96*, 1996, pp. 109–118.

Loon, J. C., Cartographic Generalization of Digital Terrain Models, PhD dissertation, The Ohio State University, University Microfilm International, Ann Arbor, MI, 1978.

Luebke D., Reddy, M., Cohen, J., Varshney, A., Watson, B., and Huebner, R., *Level of Detail for 3D Graphics*, Morgan Kaufmann, San Francisco, 2003.

Thiemann, F., Generalization of 3D building data, in *Proceedings of ISPRS Commission II Symposium "Geospatial Theory, Processing and. Applications,"* Ottawa, Canada, 2002. (*International Archives of Photogrammetry and Remote Sensing*, 34[Part 4]) (CD-Rom)

Tobeler, W., Numerical Map Generalization, Discussion Paper 8, Michigan Inter-University Community of Mathematical Geographers, 1966. Reprinted in *Cartographica*, 26(1), 9–25, 1989.

Weibel, R., An adaptive methodology for automated relief generalization, in *Proceedings of Auto-Carto 8*, Baltimore, MD, 1987, pp. 42–49.

Weibel, R., Model and experiments for adaptive computer-assisted terrain generalization, *Cartography and Geographic Information Systems*, 19(3), 133–153, 1992.

Wu, H., Structured approach to implementing automated cartographic generalisation, in *Proceedings of ICC '97*, Vol. 1, 1997, pp. 349–356.

Epilogue

It is not an easy task to write the first book on a topic. It took me two years to put together a set of algorithms in the current form. Naturally, I felt relieved and excited after the completion of the final draft of this book, but soon I started to confess and felt obliged to write this epilogue because I have faced a number of challenges during the writing process.

The first challenge was related to the selection of the topic. I believed that, after a 40-year development, it was the time to write a complete book systematically addressing the various issues concerning multi-scale spatial representation, that is, theories, algorithms, knowledge and rule databases, quality assessment, system development, as well as applications. However, I was astonished by the complexity of the issues, the immature nature of some topics, and the difficulty of organizing the materials for a complete book. Indeed, I even failed to produce a satisfactory table of contents because of the sprawling nature of the discipline. Therefore, I stuck to my usual bottom-up strategy and decided first to write a book only about the algorithmic foundation of the discipline.

The second challenge was related to the selective omission of materials. It is understandable that a complete set of operations should be available before writing algorithms for each of them. However, there was no such set available, so I had to systematically redefine existing operations and identify more operations to form a set of essential operations, although this is not necessarily a complete set. It must be noted here that the detection of conflicts was not regarded as an essential operation for multi-scale spatial representation and is thus omitted in this text. The second problem related to the selective omission of materials was what algorithms to select for a given operation, as there might be more than one algorithm for a single operation and those algorithms may operate at different levels. In this text, only selective low-level algorithms have been described, and thus those at high- levels, that is, compound and (artificial) intelligence-based algorithms, are all left out. In other words, only some typical low-level algorithms were selected. On the other hand, efforts were made to provide some mathematical solutions to operations that lack algorithms, although such solutions may be not very effective.

The third challenge was related to the depth of the discussion. Readers who are interested in mathematics may feel disappointed because neither mathematical details nor mathematical proofs have been presented. I must apologize for that and would suggest these readers trace the references. Indeed, it is my intension to write this text in *Scientific American* (i.e., semi-technical) style to enhance its readability. An attempt was also made to use more illustrations, so as to make the working principles of algorithms intuitive.

This book provides a rather comprehensive coverage of the algorithmic foundation for multi-scale representation of spatial data. I am pleased with the compilation of some materials presented to you but also feel guilty about the imperfections. Your comments are appreciated so that we can make improvements in another edition, if possible.

Index

T - #0220 - 111024 - C0 - 234/156/14 - PB - 9780367577797 - Gloss Lamination